THE ARGYLE PATENT
And Accompanying Documents

Excerpted from |||| | ||||||||| |||| || ||| |||
I0117034

HISTORY OF THE SOMONAUK
PRESBYTERIAN CHURCH

By Jennie M. Patten

With Notes on Washington County Families

CLEARFIELD

Reprinted for
Clearfield Company, Inc. by
Genealogical Publishing Co., Inc.
Baltimore, Maryland
1991, 1999

Excerpted from *History of the Somonauk United Presbyterian*
Church near Sandwich, DeKalb County, Illinois
Originally published: Chicago, 1928
Reprinted: Genealogical Publishing Co., Inc.
Baltimore, 1965, 1979
Library of Congress Catalogue Card Number 65-29271
International Standard Book Number 0-8063-0272-0

PREFACE

The material included in this volume was taken from the appendix to the History of the Somonauk Presbyterian Church by Jennie M. Patten, originally privately printed in Chicago in 1928, now a rare and expensive volume. Publication in this abbreviated form makes it possible to bring you this pertinent information at a reasonable price.

The Argyle Patent documents include four lists of immigrants, listed in Lancour's Bibliography of Ship Passenger Lists as No. 101. In addition, we have included the genealogical notices of these Washington County Families: McNaughton, Livingston, Savage, Gillaspie, and Clark.

Between 1738 and 1740 groups of Argyleshire families belonging to the Scotch Presbyterian Church, totaling 472 persons, were brought by Captain Campbell to the new world by invitation of the Provincial Governor of New York Colony, who offered a thousand acres of land to every adult person, and to every child who paid passage, five hundred acres. For various reasons the contract was not kept by the Governor. In 1764 a large number of colonists, led by Alexander McNaughton, succeeded in securing a grant of 47,450 acres, known as the Argyle Patent, in the township of Argyle and in parts of the towns of Fort Edward, Greenwich and Salem, in Washington County, upon which the Scotch colonists and their descendants took up their abode. The same year, a group of Scotch-Irish, some of them related to the settlers on the Argyle Patent, came from Pelham, Massachusetts, and settled near them, having secured a grant of 25,000 acres, known as the Turner Patent.

TABLE OF CONTENTS

The Argyle Patent and Documents:

THE ARGYLE PATENT AND ACCOMPANYING DOCUMENTS

Present day Americans can with difficulty realize the laborious steps by which their ancestors secured their first foothold upon the shores of this continent. To show how one by one the obstacles were overcome that stood between Laughlin Campbell's colonists and the establishment of their homes in the Crown lands granted to them in the name of King George III., in eastern New York, the documentary evidence of that struggle, continued through an entire generation, *is here for the first time brought together in print.* The gathering of these documents has taken half a lifetime to accomplish and is a distinct service for from them may be gleaned much that is of personal interest to a great number of the descendants of the original grantees. It will be noted that in the time that elapsed between the original petition and the granting of the Patent some changes in names occur due to death, marriage and other causes. While the period of waiting tried the patience of the colonists it was providential that the lands were not at once thrown open, for those of limited means could not have maintained themselves upon the new land in the period before it became productive.   C. M. M.

Document I

The first of the Campbell colonists landed in New York City on September 22, 1738. The following petition of heads of families, dated October 17, 1738, was presented to the Governor of the Province.

To the Honourable George Clark Esqr.,
    Lieutenant Govenour and Commander in Chief of the Province of New York &c.

    The Humble Petition of Alexander Montgomerie Alexander Mc-Naught (on) Peter McArthur and Daniel Carmichil in behalf of themselves & twenty Six other heads of ffamilys who came from North Britain and lately arrived in this province

    Sheweth That Your Petitioners being informed that there is a certain Tract of Land at or near the Wood Creek in the County of Albany now vested in the Crown And Your Petitioners being desirous to Sue out his Majesty's Letters Patent for Seven Thousand two hundred acres thereof In order to cultivate & improve the Same

    Your Petrs. Therefore humbly pray Your Honr. will be favourably pleased to grant to them their Heirs & Assignes His Majesties Letters Patent for the said quantity of Seven Thousand two hundred acres of the Lands aforesaid in Such proportions & in such Manner

APPENDIX

and under Such Quitt sale Conditions provisoes Limitations & Restritions as to Your Honr. & this Honble. Board Shall Seem Meet And Yr. Petrs. Shall Pray

Alexander Montgomery in behalf of
themselves & rest off ffamilys

(Endorsed)

Petn. of Alexander Montgomery &c for 7200 Pat ye Wood Creek 1738. Octr 17th read & referred & reported by Mr. Horsemanden in favor of ye Petrs. Petn. Warrt. not desired till they had taken a view of the lands. List of No. familys proposed to settle inclosed.

John McNeal proposes to bring in 4 familys on a grant of a thousand acres to him.

Ronald Campbell the same proposal as Mr. McNeal. Lands vested in ye Crown at or near the Wood Creek in ye County of Albany.

To begin at the North bounds of Saragtoga at the East Side of the river and soe to the Northward back of the Pattens as far as the Caring place and along the Caring place & the Wood Creek soe farr as the false on the sd. Creek. Saml. Campbell.

On the Petn. of Alexander Mtgomery & Compe.

report—That ye Lands be granted in such proportions & Divisions to each ffamily yat List therewith produced

Condn—that such ffamilys respectively Sho'd on or before the 1st. of June then next Settle on ye Said Lands and continue to inhabit there unless removed by fforce, and in case removed by fforce & obliged to quit their respective Settlemts. or Dwellings to return thereto again So Soon as Such fforce Sho'd be removed.

A List of ffamilies from the Island of North Britane

	CHILDREN	
Patrick McArthour & wife	6	300
Alexr Mc Arthour & wife	6	200
Duncan Mc Arthour & wife	6	300
Neill Mc Arthour & wife	5	200
Ronald Mc Dugall & wife	4	300
Allan Mc Dugall & wife	5	300
Archd. Mc Dugall & wife	3	300
Donald Carmichell & wife	5	300
Neill Mc Conn & wife	5	200
Donald Mc Cloud & wife	3	300
Alexander Mc Naught (on)	6	300
Donald Mc Eachern & wife	3	100
James Gillies & Broer	5	400
Duncan Tailor & wife	3	300
Archd. Mc Kellar & wife	3	200
Charles Mc Kellar & wife	3	200
Dudly Mc Duffie & wife	3	200

	CHILDREN	
Neill Mc Donald & wife	4	200
John Mc Kenzie & wife	7	240
George Mc Kinzie & wife	7	150
John Mc Niveen & wife	4	200
Cormig Mc Keay & wife	3	100
Duncan Gilchrist & wife	3	200
James Campbell & wife	6	200
Archibald Mc Eachern & wife	3	200
Donald Mc Millan & wife	2	100
Archd Jonstone & wife	2	100
Malcolm Mc Duffie & wife	7	150
Donald Campbell	4	400
Alexr Montgomery	8	560
		7,200

Document II

The foregoing petition not having been granted, the colonists resided elsewhere for a quarter of a century and on the 23d. of February 1763 certain of them again petitioned the Governor.

Petition by Alexander McNaughten, Neal Shaw, Ronald McDougall, Rich'd Campbell and one hundred others 23d. of February, 1763; The Report of the Council of the City of New York held the 2d. of March, 1763 and the Minute recording the granting of the Petition 21st. May, 1763.

To His EXCELLENCY the Honourable Robert Moncton Captain General and Govenour in Chief of the Province of New York and Territories thereon depending in America Vice Admiral of the same and Major General of his Majestys forces &c &c &c

THE PETITION of us the Subscribers HUMBLY SHEWETH

THAT some of your Petitioners are part of the fellow Emigrants of and others are descended from Persons now Deceased who also did emigrate with the Deceased Capt. Campbel from North Britain with design to form a Settlement in the Northern frontiers of this Province

THAT your Petitioners are informed that Donald Campbel George Campbel and James Campbel Sons of the said Capt. Campbel Have lately Preferred their humble Petition to your Excellency setting forth the General encouragement formerly given by the Government of this Province upon which the Aforesaid Emigration was founded the Obstructions that were raised against so laudible and useful a Design the Great distress and Poverty to which the Emigrants were reduced by the disappointment of their Scheme and the said Petitioners well grounded hopes of Effecting by the Assistance of Yoyr Present Petitioners and their relations in Argylshire the Settlement formerly intended on the Lands which the said Capt. Campbel had in view part whereof Sufficient for the Purpose are still Vacant. And therefore praying of your Excellency the Royal Grant to them the said Donald

9

APPENDIX

Campbel George Campbel and James Campbel and their Associates for one hunderd Thousand Acres in fee to be elected in one Tract on or near the Wood Creek between the falls of that Creek on the North and Batten Kill on the South a line Twenty Miles from Hudsons River on the East and that River the East side of Lake George and a South Line thence to Hudsons River on the West upon such Terms as your Excellency may think necessary to Prescribe.

AND Your Petitioners further beg leave to inform Your Excellency that the Government of this Province never treated the Said Captain Campbells fellow Emigrants as Dependants on him but as Principals in the Then intended Settlement as appears by the Copy of the Minutes of Council hereunto Annexed expressive of the favourable Intentions of the Government towards those Emigrants which however were Prevented from being carried into Execution by the last War which rendered a Settlement Impracticable in that part of the Country and together with the poverty of those Emigrants compelled them to abandon the Enterprise.

THAT as the late remarkable Success of his Majestys Arms in the Total reduction of Canada has removed every obstacle to a Settlement of that part of the Country your present Petitioners humbly beg leave to renew their aplication to the Government in favor of the said Emigration which your Petitioners Conceive the more necesssary as the said other Petitioners have never thought Proper to advise with Your Present Petitioners on the Subject matter of their said Petition however proper they thought it to avail themselves of a pretended Association with your present Petitioners on the subject matter thereof

THAT your Petitioners are informed that the said other Petitioners relying on the merit of their said Father which was common with that of his fellow Emigrants flatter themselves with an expectation of appropriating so large a proportion of the Lands which they Petition for as would not Leave a Quantity of Good Lands sufficient for the Encouragement of those whom they are pleased to call their Associates

That tho it is not the design of your Petitioners to endeavour at Obstructing the Bounty of the Crown in favor of the said other Petitioners yet it is humbly conceived that the very reasons offered by the latter will have weight to prevent any grant in their favor which would be inconsistent with the General Interest of those whom they call their Associates.

THAT many of your Present Petitioners and some of the Other Emigrants and their families tho they long felt the ill Effects of former Disappointments are at length by the Smiles of Providence on their honest endeavours not only capacitated to make larger Settlements for themselves than were Originally intended for each of the said Emigrants but are also able and willing to Assist the Others of their fellow Emigrants and their families whose Circumstances will require Aid in the Execution of the General plan Besides which as the fam-

10

ilies of many of the Emigrants consist of Several Persons Grown to Maturity your Petitioners conceive that each parcel of One Thousand Acres will Speedily be Cantoned out into several farms and the Country thereby most effectually Settled

YOUR PETITIONERS therefore most humbly pray your Excellency that the Royal Grant may issue either to your Present Petitioners and their other Emigrants and their families seperately or in Conjunction with the said other Petitioners for One Thousand Acres of Land in fee to each of your present Petitioners and the Other Emigrants now living and the families of those who are dead from and out of the Tract Above Mentioned and described in one entire Parcel within the Bounds following that is to say Beginning at or about the head of South Bay Extending Southerly to the Lands Petitioned for by Ebenezer Lacey and his Associates to extend Eastwardly towards New Hampshire Line and Westerly by Mountains and Vacant Lands still vested in the Crown and that on such Terms as your Excellency in Your Superior Wisdom may think necessary to Prescribe

AND Your Petitioners shall ever Pray &c.

City of New York

23d. of February 1763. [List of names omitted.]

At a Council held at Fort George in the City of New York the 17th. October 1738

PRESENT

The Honourable George Clarke Esqr. Lieutenant Govenour

Doctor Colden	}	Mr. Chief Justice
Mr. Livingston	}	Mr. Ccurtlandt
Mr. Kennedy	}	Mr. Horsmanden

The Petition of Alexander Montgomery, Alexander McNaught(on) Peter McArthur and Daniel Carmichel in behalf of themselves and twenty six other heads of families who came from North Britain and lately arrived in this Province was presented to the Board and Read Setting forth that the Petitioners were informed that there was a Certain Tract of Land at or near the Wood Creek in the County of Albany vested in the Crown The Petitioners therefore Prayed his Majestys Letters Patent for Seven thousand two hundred Acres thereof in such Proportions and Divisions and in Such manner as to this Honourable Board Should Seem fit

WHICH Petition having been read was referred to the Gentlemen of the Council or any five of them

HIS Honour withdrawing the Council resolved into a Committee to consider of the aforegoing Petition

THE Committee being agreed on their report by them to be made thereon and his Honour acquainted therewith.

HIS Honour returned to the Council Chamber and took his Seat Ordered that the said report be made immediately

11

Then Mr. Horsmanden Chairman of the Committee to whom the said Petition was referred in his place reported that the Committee had duly weighed and Considered of the same and as to the Petition of the aforesaid Alexander Montgomerie and Company the Committe were of Opinion that his Honour do Grant to the Petitioners the Quantity of lands by them prayed for in such Proportions and Divisions of each family as in a List therewith are Particularly mentioned and under the Conditions that such families respectively shall on or before the first day of June next Settle on the said Lands and continue to inhabit there unless removed by force and in Case the Petitioners their heirs or Assigns or any of them Shall at any time be forced by any Enemy or otherwise to go off the land and Quit their respective Settlements or Dwellings they shall return thereto again and inhabit there as soon as such Force shall be removed and that they can Inhabit there with Safety

WHICH report on the Question being put was agreed to and approved of and this Board does humbly advise and Consent that his Honour do grant to the said Petitioners his Majestys Letters Patent for the Lands by them prayed for with the Conditions and Provisions above mentioned.

(Endorsed:)

To his Excellency the Honourable Robert Monckton Captain General & Governour in Chief of the Province of New York &ca &ca &ca

THE PETITION of a Number of the fellow Emigrants of Capt. Laughlen Campbell & the descendants of others of his fellow Emigrants praying a grant of Lands in the Northern parts of this Province at the place formerly intended for their Settlement.

2d. March 1763 Read and referred to a Committee.

21st. May. Reported and granted, & Warrant of survey issued dated 21 May 1763.

G. W. Banyar D. W. Con.

Document III

The following is the report of the Committee of the Council upon the foregoing.

At a Committee of his Majesty's Council of the Province of New York held at Fort George in the City of New York the Second Day of May 1763.

PRESENT

Mr. Horsmanden		Mr. Walton
Mr. Smith	}	Mr. DeLancey
Mr. Watts		Earl of Stirling

May it Please your Excellency.

IN OBEDIENCE to your Excellency's Order in Council of the second day of March last, Referring to us the Petition of Alexander Mc-

THE ARGYLE DOCUMENTS

Nachten and others, to the Number of One hundred and seven persons, some of them Emigrants, and others of them Descendants from Persons now deceased, who emigrated with Captain Lauchlin Campbell from North Britain in the years 1738, 1739 and 1740, on the Encouragement given by Colonel Cosby Governor of this Province, in Certain proposals made and published by him with the advice of the Council in the year 1734, for the settlement of the Northern Frontier by the Protestants from Europe—praying a Grant of One thousand Acres of Land to each of the Petitioners, and the other Emigrants now living, and the Families of those who are Dead, to be laid out in one entire Tract, Beginning at or about the Head of South Bay, Extending Southerly to the lands petitioned for by Ebenezer Lacey and his Associates, to Extend Eastwardly towards New Hampshire Line, and Westerly by Mountains and vacant Lands still vested in the Crown—The Committee have considered and duely weighed the said Petition, and having made the fullest Enquiry, they could, as to the Persons now living who so emigrated, or the Descendants of those who are deceased; are humbly of Opinion that your Excellency do by his Majesty's Letters Patent grant to the Persons hereafter named the Quantity of Forty seven thousand four hundred and Fifty Acres of Land, to be laid out in one Tract, vested in the Crown lying on the East side of Hudson's River, within the County of Albany, adjoining on the South to the Ten thousand Acres of Land proposed to be Granted to Donald Campbell and others and Batten Kill; On the West to the Lands granted to John Schuyler and others on the East to the Lands proposed to be Granted to Alexander Turner and others; and to extend so far Northward as to Contain the full Quantity above Expressed. That the same be granted on the Quit Rent Provisoes Limitations and Restrictions prescribed by his Majesty's Instructions, and that the said Grant be made to Duncan Read Alexander Montgomery Alexander McNachten, Neal Shaw, Henry Van Vleck, Archibald Campbell, George Campbell, Neal Gillaspie, Alexander McLean and Ennis Graham and their Heirs, as Trustees To hold the Quantity of five hundred Acrse, part of the said Larger Tract so to be Granted as aforesaid, in Trust to and for the use of a Minister and Schoolmaster resident on the said larger tract for ever, AND to hold all the Residue and Remainder of the said Larger Tract, in Trust to and for the Respective uses of the several persons named in the Schedule Hereunto annexed, and their Heirs, in the Proportions in and by the said Schedule alloted to the said persons respectively. AND that the whole of the said Tract of Land be erected into a Township with the usual priviliges by the Name of Argyle.

All which is nevertheless humbly submitted

New York
2d May 1763.
(Endorsed)

By Order of the Committee.
Wm. Smith Chairman.

13

APPENDIX

Report of the Committee on the Petition of Alexander McNachten and others.

SCHEDULE

	ACRES		ACRES
Alexander Montgomery	600	Barbary McAlister	300
Daniel Johnson	350	Jannet Ferguson	250
Elizabeth McNeil	300	William Clark	350
Archibald Campbell Senr.	300	Issabela Livingston	250
John McCarter	400	John McEuen	500
John Shaw Senr.	300	James Campbell	300
James Gilles	500	Duncan McDuffie	350
Duncan Taylor	600	Allan McDonald	300
Donald McMullen	500	Duncan Read	600
Mary McCloud	250	John Shaw Junr.	300
Edward McCay	300	Neil Shaw	600
Ronald McDougall	400	Archibald McGowne	300
John McDougall	400	John McGowne Junr.	250
Archibald McDougall	450	John McGowne Senr.	300
Dougall McCaller	550	Donald McMullen	450
Edward McCaller	500	Ann Duffie	350
Alexander McNauchten	600	Duncan McGuire	500
Archibald McNiven	350	Duncan Lindsay	350
Patrick McArthur	350	Neil Carmichel	300
John McCarter	350	John Read	450
Duncan McCarter	450	Neil Carmichel	300
Neil McEachron	450	Duncan McDougall	500
Neil McDonald	500	Archibald Campbell Junr.	250
Duncan Gilchrist	500	John McFail	300
Florence McKinzie	200	Archibald McCollum Senr.	350
George McKinzie	400	John McIntire	350
Malcolm McDuffie	550	Marian Campbell	250
John McDuffie	250	Duncan Campbell Senr.	450
Dougall McAlpine	300	Alexander Christie	350
Robert Campbell Senr.	350	Alexander McArthur	250
William Fraser	350	Daniel Clark	350
Hannah McEhen	400	Daniel Shaw	350
James Nutt	300	Hugh McIlvray	200
Elizabeth Cal	250	Dougall McDuffie	350
Neil McPhaden	300	Duncan Mc Phaden	300
John McGuire	400	Archibald McCollum Junr.	350
Catharine McCarter	200	David Torry	300
Dougall Thompson	400	William Hunter	300
Mary Anderson	300	John McArthur	300
Robert Campbell Junr.	450	John McCollum	300
Charles McAlister	300	Duncan Mc Kinvan	350

14

THE ARGYLE DOCUMENTS

	ACRES		ACRES
Mary Anderson	300	Margaret Cargill	250
Hugh Montgomery	300	Ann McArthur Senr.	250
Mary Beton	300	Jane Widrow	300
Alexander McDonald	250	John Campbell	300
Mary Graham	300	Mary Hammels	250
William Graham	300	Margaret McAlister	250
Hugh McDougall	300	Angus Graham	300
Angus McDougall	300	Roger McNeal	300
Rachel McNiven	300	Anna McArthur	300
John Gilchrist	300	Margaret Gilchrist	250
Alexander Gilchrist	300	John Torry	300
Donald McIntire	350	John McCore	300
Catharine McLean	300	Archibald McCore	300
Archibald McEuen	300	Charles McArthur	350
Catharine Campbell	250	James McDonald	350
Jane Cargyle	250	Alexander Campbell	350
Florence McVarick	300	George Campbell	300
Catharine Shaw	250	Duncan Shaw	300
Archibald McIlfender	300	Alexander McDougall	350
Catharine McIlfender	250	Eleanor Thompson Widow	
Roger Reed	300	of Roger Thompson	300
Mary Torry	250	Hugh McCarty	300
Angus McDougall	300	Neal McEuen's Daughter	
Malcolm Campbell	300	Marian McEuen	200
Mary Campbell	250	Elizabeth Frazier	200
Robert McAlpine	300	Elizabeth Ray	200
Duncan Campbell Junr.	300	Parsonage and School	500
Duncan Campbell the third	300		
Elizabeth Campbell	300		6250
Ann Campbell	300		11300
John McIntire	300		14250
Elizabeth Cargill	250		15650
James Cargill	300		
John Cargill	300		47450

The above is the Schedule Referred to in the Report of the Committee on the Petition of Alexander McNachten and others of the 2d day of May, 1763.

By Order of the Committee.   William Smith, Chairman.

(Endorsed) 2d May, 1763:

Report of the Committee on the Petition of Alexander McNachten and others, 21st May, 1763. Read and Confirmed. No. 4. Entered.

15

APPENDIX

Document IV

Memorial of Duncan Read & four others praying that the Memorialists may be the sole Trustees in the Grantes ordered to the Persons who emigrated with Captain Lauchlin Campbell deceased. 14, Sepr. 1763. Read in Council.

To the Honourable Cadwallader Colden Esqr. His Majesty's Lieutenant Governor and Commander in Chief of the Province of New York and the Teritories Depending Thereon in America.

In Council

THE MEMORIAL of Duncan Reid, Neil Shaw, Archid. Campbell, Alexander Mc Nachten & Neil Gillaspie five of the Patentees Nominated for the Tract of land Surveyed and Laid out for Sundry Scotch People who Emigrated with Captain Lauchlin Campbell.

SHEWETH THAT your Memorialist find themselves utterly unable to Collect or Raise the Monies Necessary towards the obtaining of the Patent from the Several Persons who have Shares therein, many of them (altho' very desirous of having the Land) yet (on Various Pretents) Refuse Paying any money whatsoever while Others Insist that they will not Pay any until the Patent is Actually Issued and those who are willing to Pay Their Proportions, keep back the Money in Justice to themselves, until the others Comply.

THAT your memorialists and some Others of the Patentees were about Raising a sum Sufficient for this Purpose upon Interest, and had Actually Procured the same but some of them Refseing to Execute the Bond for the Repayment Thereof, this Method was Drop't so that your Memorialists are utterly at a Loss what Measures to Pursue, as Delays in suing out the Pattent will be Attened with the worst Consequence to your Memorialists and those others who are Actually Prepared to Settle in May next having for that Purpose Delivered up the farms they now live on to Their Landlords, so that your memorialists would humbly Suggest to your honor that the Names of such Persons who are Present Nominated as Patentees and who Refuse Joining in Raising Money for this Necessary Purpose be Struck out and the Number Confined to your Memorialists who will undertake to Raise such sums of Money as shall be Necessary, and that a Clause May be added in the Pattent for the Security of your Memorialists (who must Otherwise run too great a Risque) that they shall not be Obliged to Convey to Their Associates the respective Quantities of Land Alloted to them before Each of Them do first Pay or Secure their Just Proportion of the fees and Expenses which shall Attend the Obtaining the Patent, this Method your Memorialists humbly Conceive would Remove all Difficulties.

16

AND Therefore humbly Prays that the same may be adopted or Such Other Measure as to your Honor Shall Appear to have a greater Tendency to facilitate the obtaining Of the Patent

And your Memorialist will ever pray
New York Septr. 14th. 1763.

> Duncan Read
> Neal Shaw
> Archid. Campbell
> his
> Neil X Gillaspie
> mark
> Alexr. Mc Nachten

Document V

Petition dated February 1st, 1764 to the Honourable Cadwallader Colden Esqr. his Majesty's Lieut. Govenour, and Commander in Chief of the Province of New York, and the Territories depending thereon in America. IN COUNCIL

The Humble Petition of Duncan Read, Neal Shaw, Archibald Campbell, Alexander McNachten & Neal Gillaspie, Trustees for the Emigrants of Captain Lauchlin Campbell,

HUMBLY SHEWETH

THAT your Petitioners run great Hazard, in becoming Security for all Such Costs, and Charges, as shall Attend the Obtaining his Majesty's Letters Patent, for the Lands Intended to be Granted, subdividing the said Lands & Conveying the same in the Proportions alloted to each Party, and that no Method is Laid out whereby they Can Reimburse themselves the Money they shall Expend, in Cases where any of the Parties shall Delay Neglect or refuse to Pay his Proportion of the costs and Charges.

Your Petitioners Therefore humbly Pray that a Clause may be Inserted in the Letters Patent, Empowering your Petitioners, or the Survivor or Survivors of them, to sell and Dispose of the shares of such Grantees, as do not within such Certain time as to your Honour shall seem Reasonable, Pay unto your Petitioners all such Expences as shall Attend the Execution of Their Trust.

AND your Petitioners will Pray &c;

> Duncan Read
> Neal Shaw

	ACRES
Alexander Montgomery	600
John McNeil's four Daughters	300
Ann McDougall, (Campbell) Archibald the Son & Isabell the Daughter	300
Neil McArthurs Widdow & 5 Childrn	400

17

APPENDIX

ACRES

	ACRES
Donald Shaws Son and Daughter	300
Elizabeth Sutherland & four Childn	500
Duncan Taylor his Wife & 8 Childn	600
Donald McMillen Wife and five Childn	500
Donald McClouds Daughter	250
Cormack McCoy, Widdow Son & Daughter	300
Ronald McDougall, Wife & 3 Childn	400
Allan McDougalls, Widdow & 5 Childn	400
Archibald McDougall, Wife and 5 Childn	450
Archibald McKellers, Widdow & 8 Child	550
Charles McKellers, Widdow & 7 Childn	500
Alexander McNaught(on) Wife 4 Children and 8 Grand	600
John McNevin Dead left 1 son and four Daughters	350
Patrick McArthur, Wife 2 Sons and One Daughter	350
Duncan McArthur's Widdow 2 Sons and 2 Daughters	350
Alexadder McArthurs 7 childn	450
Donald McEachern Widdow & Six Children	450
Neil McDonald Wife & 6 Children	500
Duncan Gilchrist Wife & 6 Children	500
John McKinzie's Daughter	200
Geo. McKinzie Wife & four Childn	400
Malcolm McDuffie Wife & 7 Children	550
Dudley McDuffie's 2 Children	250
Dugal McAlpine Wife and 2 Children	300
Donald Campbell Widow 4 Children	350
Robert Fraziers four Children	350
Archibald McEuen Dead left 2 Childn	250
James Nut and Son	300
John Colwell's Daughter Widow Martin	250
Neil McPhadon Wife and Daughter	300
John McGuire Wife and 4 Children	400
Patrick McArcher's (Eacherns) Widdow	200
Dugal Thompson Wife three Sons & Neice	400
Patrick Anderson's Widdow and two Daughters	300
Duncan Campbells Widdow 3 Sons and One Daughter	350
Charles McAlister's 2 Sons	300
Duncan McAlister's 1 Son & 2 Dau'rs	300
Donald Ferguson's Daughter & Neice	250
William Clarke Wife Son & Daughter	350
Donald Livingston's Widdow & Daur	250
John Mc Euen Wife and five Sons	500
Murdoc McInnish Descendants	300
Archibald McDuffie Dead 1 son	250
Neil McEnnish (McInnish) Widdow married to Alan McDonald	200

18

ACRES

Duncan Reid brought a Wife and 8 Children	600
Neil Shaws Grand Children	300
Neil Shaw the Eldest	300
Archibald McGowne 2 Children	300
Malcolm McGowne Dead 1 son	250
John McGowne and Wife	300
Donald McMillen, Wife and 5 Childn	450
Archdd. McDuffiie Widdow & 2 Daurs &c	350
Duncan McGuire Wife & 5 Childn. &c	500
Donald Lindsey 1 Son 2 Daurs. &c	350
Neil Galaspie Wife 2 Sons & one Daur	450
John Reid Wife and five Children	450
Dugal Carmichael Dead, 1 son	300
Duncan McDougall Wife & five Childn	500
Archibald the son of Duncan Campbell	250
John Mc Fail Widdow Son and Daur	300
Archibald McCollom 2 Sons 1 Daur	350
Nicholas McEntire Widdow 2 Sons and 2 Daughters	350
James Storie Dead 4 Children	300
Alexander Hunter Dead Son and Daur	300
Alexander McArthur Widdow & 1 son	300
Alexander Campbell Dead 2 Daurs	250
Duncan Campbell Wife & four Childn	450
John Christies Widdow and four Childn	350
John McArthur one Son and one Daur	250
Angus Clarke 2 sons & 1 Daur & Grand Children	350
John Shaws Widdow and four Childn	350
John Mc Elery (McGillivray	200
Dudley McDuffies Widdow 2 Sons & 2 Daughters	350
Duncan Mcphadons 2 Sons	300
Archibald McColloms son & Daur	350
Archibald McColemans Widdow One son and two Daughters	300
Duncan McKinven and four Childn	350
Mary Anderson and two Daughters	300
Hugh Montgomery	300
Mary Beacon	300
Jennet Fergusons Son	250
Mary Grahams Children	300
Alexander Grahams two Sons	300
Hugh McDougall	300
Marian McNevan	300
Rachel McNevan	300
John Gilchrist	300
Alexander Gilchrist	300

19

	ACRES
Donald McIntire	350
Lauchlin McLeans Daur. Cath:	300
Malcolm McEuens 3 Children	300
Catharine Campbell	250
Jane Cargyl now Mrs. Van Vleet	250
Florence McVarrick	300
Catharine Shaw	250
Archibald McIlfender	300
Roger Reed	300
George Storys Child	250
Angus McDougall	300
David Shaws Widdow	250
Malcolm Campbell	300
Archibald Campbells Daur.	250
Robert McAlpine	300
Duncan Campbell	300
William Campbells Childn.	300
Archibald Campbells Childn	300
Catharine McIlfender	250
Anne Campbell	300
John McIntire of Pensilvana	300
Elizabeth Cargill	250
James Cargyl	300
John Cargyl	300
Margt. Cargyl	250
Anna McArthur	250
Jane Widrow	300
John Campbell	300
Mary Hammels Daughter	250
Margaret Mcalister	250
Angus Graham	300
Roger McNeal	300
Anna McArthur	300
Margaret McGilchrist	250
John Tory	300
For a Parsonage & School	500
John McCore	300
Archibald McCore	300
Charles McArthur	350
Alexander McDonald	300
Alexander Campbell	350
George Campbell Son of John Campbell	300
Neal and Duncan Shaw, Sons of John Shaw	300
Alexander McDougall	350

THE ARGYLE DOCUMENTS

Document VI

THE ARGYLE PATENT

GEORGE the Third by the Grace of God of Great Britain France and Ireland King Defender of the Faith and so forth—To ALL to whom these Presents shall come GREETING WHEREAS Alexander MacNachten and others our Loving Subjects, to the Number of one hundred and seven Persons in the whole, some of them Emigrants, and others of them Descendants of several Persons now deceased, who emigrated with Captain Laughlin Campbell from North Britain in our Kingdom of Great Britain, in the Years of our Lord One thousand seven hundred and thirty Eight, One thousand seven hundred and thirty nine, and one thousand seven hundred and forty; with Design to Form a Settlement on the northern Frontier of our Province of New York, on the Encouragement given by William Cosby Esquire then Governor of the said Province, in Certain proposals made and published by him, with the Advice of the Council of the said Province in the Year of our Lord One thousand seven hundred and thirty four; for the Settlement of the northern Frontiers of the said Province by Protestants from Europe, by their humble Petition presented to our trusty and well beloved Robert Monckton our Captain General and Governor in Chief, in and over our said Province of New York, and the Territories thereon depending in America, Vice Admiral of the same, and Major General of our Forces, and read in our Council for our said Province on the second day of March last, did humbly pray a Grant of one Thousand Acres of Land to each of them the Petitioners, and the others the said Emigrants now living, and the Families of those other Emigrants who are dead, to be laid out in one entire Tract, beginning at or About the Head of South Bay extending southerly to the Lands Petitioned for by Ebenezer Lacey and his Associates to extend Eastwardly towards New Hampshire Line, and westerly by Mountains and vacant Lands still vested in the Crown, on such Terms as our said last mentioned Captain General and Governor in Chief of our said Province of New York should think necessary to prescribe; which Petition having been then and there duly weighed, read and refered to a Committee of our said Council, our said Council did afterwards on the Twenty first day of May following in Pursuance of the Report of the said Committee humbly advise the same our Governor by our Letters Patent to grant the Quantity of Forty seven Thousand four Hundred and fifty Acres of Land to be laid out in one Tract vested in the Crown, lying on the East side of Hudson's River within the County of Albany adjoining on the South to the Ten Thousand Acres of Land proposed to be granted to Donald Campbell and others and Batten Kill, on the west to the Lands granted to John Schuyler and others, on the East to the Lands proposed to be granted to Alexander Turner and others, and to extend so far Northward as to

21

contain the full Quantity above expressed (on the Quit Rent, Provisoes, Limitations and restrictions prescribed by our Royal Instructions) to Duncan Reid, Alexander Montgomery, Alexander Mac Nachten, Neal Shaw, Henry Van Vleck, Archibald Campbell, George Campbell, Neal Gillaspie, Alexander Mac Lean, and Ennis Graham, and their Heirs as Trustees; to hold the Quantity of Five hundred Acres part of the said larger Tract so to be granted as aforesaid in Trust to and for the Use of a Minister and Schoolmaster Resident on the said larger Tract forever: And to hold all the Residue and Remainder of the said larger Tract in Trust to and for the respective Uses of the Several Persons named in the Schedule to the said Report annexed, and their Heirs in the Proportions in and by the said Schedule allotted to the said Persons respectively, and that the whole of the said Tract of Land should be erected into a Township by the Name of Argyle.

AND WHEREAS afterwards on the Twenty ninth Day of September now last past; on the Memorial of Duncan Reid, Neal Shaw, Archibald Campbell, Alexander Mac Nachten and Neal Gillaspie, five of the above named Trustees presented unto our trusty and well beloved Cadwallader Colden Esquire then and now our Lieutenant Governor and Commander in Chief in and over our said Province of New York and the Territories thereon depending in America, in Council, it was for the Reasons in the said Memorial assigned, ordered that the said Memorialists Duncan Reid, Neal Shaw, Archibald Campbell Alexander Mac Nachten and Neal Gillaspie should remain, and they the said Memorialists were by the said order appointed sole Trustees to whom the Grant of the said Lands should be made in Trust to and for the several Uses of the several Persons whose Names are inserted in the Schedule above mentioned in such Proportions as are therein expressed to be conveyed by the said Trustees, or the Survivors or Survivor of them to the said several Persons respectively their Heirs or Assignes, they first paying the Charge of such Conveyance and their Proportionable Part of all such Fees and Expences as the said Trustees shall be put to in obtaining this our Grant, and in making the Division of the said Lands.

WHEREFORE In Obedience to our Royal Instructions our Commissioners appointed for the setting out all Lands to be granted in our said Province, have set out for them the said Duncan Reid, Neal Shaw, Archibald Campbell, Alexander Mac Nachten and Neal Gillaspie, in Trust to and for the Uses aforesaid, ALL that certain Tract or Parcel of Land Situate lying and being on the East side of Hudson's River in the County of Albany, Beginning on the east Banck of the said River at the south west Corner of a Tract of Land granted to James Bradshaw and others, called Kingsbury; and runs thence along the south Bounds of the said Tract, East, four Hundred and ninety two Chains to the south east Corner thereof; and then along the East

22

Bounds of the said Tract called Kingsbury North four Chains Then East two Hundred and thirty six Chains; Then South Eight Hundred and Eighty two chains, to the Middle of a Stream of Water called Batten Kill; then down the Middle of the said Stream as it runs, including the half of the said Creek or Kill called Batten Kill, to the East Bounds of a Tract of Land lately surveyed for Donald Campbell and others; Then along the said East Bounds of the said Tract surveyed for Donald Campbell and others North, Three Hundred and sixty seven Chains to the north east Corner thereof, and then along the North Bounds of the same Tract West Three Hundred and seventeen Chains to the East Bounds of a Tract of Land Granted to John Scuyler Junior and others, then along the said East Bounds of the last mentioned Tract North, Nine Degrees East, six Hundred and fifty one Chains to the North East Corner of the said Tract, Then West Thirty three Chains; then South Sixty Degrees West, six Chains; to a Tract of Land Granted to Stephen Bayard; Then along the North Bounds of the last mentioned Tract, West two Hundred and five Chains to Hudson's River; Then up the Stream of the said River as it runs to the Place where this Tract first began containing Forty seven Thousand four Hundred and fifty Acres of Land and the usual Allowance for Highways; and in setting out the said Tract of Land, our said Commissioners have had regard to the Profitable and unprofitable Acres, and have taken Care that the Length thereof doth not extend along the Banks of any River otherwise than is conformable to our Royal Instructions for that Purpose, as by a Certificate thereof under their Hands bearing Date the first Day of the month of February last, and entered of Record in our Secretary's Office, may more fully appear . . . Which said Tract of Land so set out as aforesaid according to our said Royal Instructions, We being willing to Grant to them the said Duncan Reid, Neal Shaw, Archibald Campbell, Alexander Mac Nachten and Neal Gillaspie their Heirs and Assignes with the several Powers and Priviledges and to upon and for the several Uses and Trusts herein after particularly mentioned, limited and appointed of and concerning the same and of and concerning every part and parcel thereof respectively—

KNOW YE That of our Especial Grace certain Knowledge and meer Motion, we have given, granted, ratified and confirmed, and Do by these Presents for us our Heirs and Successors give, grant, ratify and confirm, unto them the said Duncan Reid, Neal Shaw, Archibald Campbell, Alexander Mac Nachten and Neal Gillaspie their Heirs and Assigns for ever; ALL That the said Tract or Parcel of Land set out, abutted, bounded and described in manner and form as above mentioned, together with all and singular the Tenements Hereditaments Emoluments and Appurtenances thereunto belonging or appertaining: And also all our Estate, Right, Title, Interest, Possession, Claim and Demand whatsoever of in and to the same Lands and

23

Premises, and every Part and Parcel thereof; and the Reversion and Reversions, Remainder and Remainders, Rents, Issues and Profits thereof, and of every Part and Parcel thereof;

EXCEPT and always reserved out of this our present Grant, unto us our Heirs and Successors for ever all Mines of Gold and Silver, and also all white or other Sorts of Pine Trees fit for Masts, of the Growth of twenty four Inches Diameter and upwards, at Twelve Inches from the Earth, for Masts for the Royal Navy of us our Heirs and Successors, To HAVE AND To HOLD the said Tract of Land, Tenements Hereditaments and Premises hereby granted ratified and confirmed, or hereby meant mentioned or Intended so to be and every Part and Parcel thereof with their and every of their appurtenances (Except as is hereinbefore excepted) unto them the said Duncan Reid, Neal Shaw, Archibald Campbell, Alexander Mac Nachten and Neal Gillaspie their Heirs and Assigns for ever, to for and upon the several and respective Use and Uses, Intents and Purposes herein after expressed limited declared and appointed of and concerning the same and of and concerning every part and Parcel thereof, and to and for no Other Use or Uses Intents or Purposes whatsoever; That is to say,

To HAVE AND To HOLD The Quantity of Six hundred Acres together with the Usual Allowance for Highways, Part and Parcel of the said Tract of Land and Premises hereby granted ratified and confirmed With the Appurtenances to the same respectively belonging (Except as is herein before excepted) In Trust to and for each of the Persons severally herein next undermentioned (being Persons named in the Schedule herein before mentioned) their Heirs and Assignes, that is to say, to and for the only Proper and Seperate Use and Behoof of Alexander Montgomery, Duncan Taylor, Alexander Mac Nachten, Duncan Read and Neal Shaw, and each of them their and each of their Heirs and Assignes respectively forever, and to and for no other Use or Uses Intent or Purpose whatsoever

AND TO HAVE AND TO HOLD the Quantity of Five Hundred and fifty Acres together with the usual Allowance for Highways, also Part and Parcel of the said Tract of Land and Premises, hereby granted ratified and confirmed, with the Appurtenances to the same respectively belonging (except as is herein before excepted) in Trust to and for each of the Persons severally herein next undermentioned (being also Persons named in the Schedule herein before mentioned) their Heirs and Assignes, that is to say; To and for the only proper and separate Use and Behoof of Dougall Mac Caller and Malcolm Mac Duffie and each of them their and each of their Heirs and Assignes respectively for ever, and to and for no other Use or Uses Intent or Purpose whatsoever

AND TO HAVE AND TO HOLD the Quantity of Five Hundred Acres, together with the usual Allowance for Highways also Part and

24

Parcel of the said Tract of Land and Premises hereby granted ratified
and confirmed with the Appurtenances to the same respectively belonging (except as is herein before excepted) in Trust to and for each
of the Persons severally herein next undermentioned (being also Persons named in the Schedule herein before mentioned) their Heirs and
Assignes, that is to say, to and for the only proper and separate Use
and Behoof of James Gilles, Edward Mac Caller, Neil Mac Donald,
Duncan Gilchrist, John Mac Euen, Duncan Mac Guire and Duncan
Mac Dougal and each of them, their and each of their Heirs and
Assignes respectively forever, and to and for no other Use or Uses
Intent or Purpose whatsoever

AND TO HAVE AND TO HOLD the Quantity of Four Hundred
and fifty Acres, together with the usual Allowance for Highways also
Part and Parcel of the said Tract of Land and Premises hereby granted
ratified and confirmed with the Appurtenances to the same respectively
belonging (except as is hereon before excepted) In Trust to and for
each of the Persons severally herein next unmentioned (being also
Persons named in the Schedule herein before mentioned) their Heirs
and Assignes, that is to say, to and for the only proper proper and
seperate Use and Behoof of Archibald Mac Dougal, Duncan Mac
Carter, Neal Mac Eachran, Robert Campbell Junior, Donald Mac
Mullen, Niel Gillespie, John Read, and Duncan Campbell Senior, and
each of them their and each of their Heirs and Assignes respectively
for ever, and to and for no other Use or Uses Intent or Purpose
whatsoever

AND TO HAVE AND TO HOLD the Quantity of four Hundred
Acres together with the usual Allowance for Highways, also Part and
Parcel of the said Tract of Land and Premises hereby granted ratified
and confirmed with the Appurtenances to the same respectively belonging (except as is herein before excepted) In Trust to and for each of
the Persons severally herein next undermentioned (being also Persons
named in the Schedule herein before mentioned) their Heirs and
Assigns, that is to say, To and for the only Proper and seperate Use
and Behoof of John Mac Carter, Ronald Mac Dougal, John Mac
Dougal, George Mac Kinzie, Hannah Mac Euen, John Mac Guire and
Dougal Thomson, and each of theim their and each of their Heirs
and Assigns respectively for ever, and to and for no other Use or Uses
intent or Purpose whatsoever

AND TO HAVE AND TO HOLD, the Quantity of Three Hundred
and fifty Acres, together with the usual Allowance for Highways, also
Part and Parcel of the said Tract of Land and Premises hereby granted
ratified and confirmed with the Appurtenances to the same respectively
belonging (except as is herein before excepted) in Trust to and for
each of Persons severally herein next undermentioned (being also
Persons named in the Schedule herein before mentioned) their Heirs
and Assigns: That is to say to and for the only proper and seperate

25

APPENDIX

Use and Behoof of Daniel Johnson, Archibald Mac Niven, Patrick
Mac Arthur, John Mac Carter, Mary Campbell, William Fraser, William Clark, Duncan Mac Duffie, Ann Duffie, Duncan Lindsay, Archibald Mac Collom Senior, John Mac Intire, Alexander Christie, Daniel
Clark, Daniel Shaw, Dougal Mac Duffie, Archibald Mac Collum
Junior, Duncan Mac Kinvan, Donald Mac Intire Charles Mac Arthur,
James Mac Donald, Alexander Campbell and Alexander Mac Dougal,
and each of them their and each of their Heirs and Assignes respectively forever, and to and for no other Use or Uses Intent or Purpose
whatsoever; and

To HAVE AND TO HOLD the Quantity of Three Hundred Acres
together with the usual Allowance for Highways, also Part and Parcel
of the said Tract of Land and Premises hereby granted ratified and
confirmed with the Appurtenances to the same respectively belonging
(Except as is herein before excepted) In Trust to and for each of
the Persons severally herein next undermentioned (being also Persons
named in the Schedule herein above mentioned) their Heirs and As-
signes That is to say to and for the only Proper and seperate Use and
Behoof of Elizabeth Mac Neil, Archibald Campbell Senior, John Shaw
Senior, Edward Mac Coy, Dougal Mac Alpine, James Nutt, Neil Mac
Phaden, Mary Anderson, Charles Mac Allister, Barbara Mac Allister,
James Campbell, Allan Mac Donald, John Shaw junior, Archibald
Mac Gowne, John Mac Gowne senior, Neil Carmichel, John Mac Fail,
David Torry, William Hunter, John Mac Arthur, Duncan Mac
Phaden, John Mac Collman, Mary Anderson, Hugh Montgomery,
Mary Beton, Mary Graham, William Graham, Hugh Mac Dougal,
Angus Mac Dougal, Rachel Mac Niven, John Gilchrist, Alexander Gil-
christ, Catharine Mac Lean, Archibald Mac Euen, Florence Mac
Varick, Archibald Mac Ilfender, Roger Reed, Angus Mac Dougal,
Malcolm Campbell, Robert Mac Alpine, Duncan Campbell Junior,
Duncan Campbell the Third, Elizabeth Campbell, Ann Campbell, John
Mac Intire, James Cargill, John Cargill, Jane Widrow, John Campbell,
Angus Graham, Roger Mac Neal, Anna Mac Arthur, John Torry,
John Mac Core, Archibald Mac Core, George Campbell Duncan Shaw,
Eleanor Thompson, (widow of Roger Thompson) and Hugh Mac
Carty, and each of them, Their and each of their Heirs and Assignes
respectively forever, and to and for no other Use or Uses Intent or
Purpose whatsoever

AND TO HAVE AND TO HOLD the Quantity of Two Hundred
and Fifty Acres together with the usual Allowance for Highways, also
Part and Parcel of the said Tract of Land and Premises hereby granted
ratified and confirmed with the Appurtenances to the same respectively
belonging (Except as is herein before excepted) In Trust to and for
each of the Persons severally herein next undermentioned (being also
Persons named in the Schedule herein before mentioned) their Heirs
and Assignes: That is to say, to and for the only Proper and Seperate

26

THE ARGYLE DOCUMENTS

Use and Behoof of Daniel Lindsay, Margret Mac Neil, Mary Mac Cloud, John Mac Duffie, Elizabeth Calwell, Jannet Ferguson, Isabella Livingston, John Mac Gowne Junior, Archibald Campbell Junior, Marian Campbell, Alexander Mac Arthur, Alexander Mac Donald Catherine Campbell, Jane Cargyle, Catherine Shaw, Catherine Mac Ilfender, Mary Torry, Mary Campbell, Elizabeth Cargill, Margaret Cargill, Ann Mac Arthur Senior, Mary Hammels, Margaret Mac Allister, and Margaret Gilchrist and each of them their and each of their Heirs and Assignes respectively forever, and to and for no other Use or Uses Intent or Purpose whatsoever

AND TO HAVE AND TO HOLD the Quantity of two Hundred Acres together with the usual Allowance for Highways, also part and parcel of the said Tract of Land and Premises hereby granted ratified and confirmed, with the Appurtenances to the same respectively belonging (except as is herein before excepted) In Trust to and for each of the Persons severally herein next undermentioned (being also Persons named in the Schedule herein before mentioned) their Heirs and Assignes that is to say to and for the only proper and seperate Use and Behoof of Florence Mac Kinzie, Catherine Mac Carter, Hugh Mac Elvray, Marian Mac Euen Daughter of Neal Mac Euen, Elizabeth Frazier, and Elizabeth Roy and each of them, their and each of their Heirs and Assignes respectively for ever, and to and for no other Use or Uses Intent or Purpose whatsoever, Which same several smaller Tracts and Quantities of Land, the Uses and Trusts whereof are herein before respectively limited appointed and declared as aforesaid, amount to forty six Thousand Nine Hundred and Fifty Acres of Land Part and Parcel of the Larger Tract of Forty seven Thousand four Hundred and fifty Acres of Land by these Presents Granted ratified and confirmed as aforesaid And as for and concerning the remaining five Hundred Acres with the Usual Allowance for Highways of the said Tract of Forty seven Thousand four Hundred and fifty Acres of Land

To HAVE AND TO HOLD the same with the Appurtenances thereunto belonging In Trust as a Glebe forever to and for the Use Benefit and Behoof of the Minister of the Gospel and Schoolmaster for the Time being resident and officiating on the said larger Tract of Land hereby granted, and to and for no other Use or Uses Intent or Purpose whatsoever: All and singular the said Tract of Land abutted bounded and described in manner above mentioned and Premises,

To BE HOLDEN of us our Heirs and Successors in free and common Socage as of our Manor of East Greenwich in our County of Kent within our Kingdom of Great Britain Yielding rendring and paying therefore yearly and every Year for ever unto us our Heirs and Successors, at our Custom House in our City of New York unto our or their Collector or Receiver General there for the Time being, on the Feast of the Annunciation of the Blessed Virgin Mary commonly called Lady Day, the yearly Rent of two Shillings and six Pence

27

APPENDIX

sterling for each and every hundred Acres of the above granted Lands
and so in Proportion for any lesser Quantity thereof, saving and except
for such Part of the said Lands allowed for Highways as above mentioned, in Lieu and Stead of all other Rents, Services, Dues, Duties and
Demands whatsoever for the hereby granted Lands and Premises, or
any Part thereof

AND for the more easy just and equitable determining, ascertaining
and locating in the larger Tract of Land, by these Presents granted
ratified and confirmed, the Situation and Place of each of the smaller
Tracts or Lots of Land for which the Uses and Trusts have been herein
before respectively declared limited and appointed it is our Royal Will
and Pleasure that the same shall respectively be ascertained and deter-
mined by Ballot: And for that Purpose we direct and appoint that
the Trustees herein before nominated, their Heirs or Assigns or the
Survivors or Survivor of the said Trustees, their or his Heirs or
Assignes shall within one Month Next ensuing the Date hereof appoint
some convenient Time and Place within our said Province for the
Balloting aforesaid, and shall give Publick Notice thereof in all the
Publick Newspapers of this Province for four Weeks successively, and
on the Day and at the Place so appointed and notified every the Person
and Persons to whose Use and Behoof respectively the said several
Parcels or Quantities of Land are by these Presents held in Trust by
the Trustees as aforesaid or their Heirs or Assignes, shall by themselves,
or their Attorneys respectively draw Lots for the Place where the several and respective Quantities of Land aforesaid so holden for them
respectively in Trust as aforesaid shall on the said larger Tract of
Land hereby granted be located fixed and measured out to him or her:
which same Drawing shall continue there from Day to Day until the
whole is finished: And in case any of the said Persons their Heirs or
Assignes for whose Use any of the said Quantities of Land are respectively held in Trust as aforesaid, shall fail to appear by themselves
or their Attorneys respectively on the Balloting aforesaid: Then our
Will and Pleasure is That the said Trustees or either of them their
Heirs or Assignes shall draw Lots for such of them as shall so fail
to appear, and such Location as shall so be drawn for them respectively
shall be the Location of their several and respective Shares and Quantities of the said Land AND our Will and Pleasure further is, and
we do by these Presents further declare and appoint That as soon after
the Balloting herein before directed as conveyniently can be; The
Trustees herein before mentioned, their Heirs or Assignes or the
Survivors or Survivor of them their or his Heirs or Assignes shall
cause the larger Tract of Land hereby granted to be surveyed and
divided and the several and respective Quantities of Land aforesaid
Parts of the said larger Tract to be respectively located laid out and
measured for the Persons for whom they are respectively held in
Trust as aforesaid by these Presents in the Place and Places in the said

28

larger Tract where on the Ballotting aforesaid the Share or Shares of such Person shall be fixed and ascertained by Lot as aforesaid, and shall convey the same with the Appurtenances by good and sufficient Assurances in the Law to the said Persons respectively their respective Heirs and Assigns or to such other Person or Persons as they shall respectively nominate and appoint their Heirs and Assigns respectively forever. To their only proper and seperate Use and Behoof respectively forever. He or they respectively paying the Charge and Costs of such Conveyances and their Proportionable Part of all such Fees and Expences as the said Trustees their Heirs or Assigns or the Survivors or Survivor of them his or their Heirs or Assigns shall be put to and expend in and for the obtaining this our Grant, and in making the division of the said Lands with lawfull Interest for the same, from the respective Times of disbursing and expending the same and every respective Part thereof, the said Proportions to be struck according to the Quantities of Acres in each share respectively

PROVIDED ALWAYS and it is our Royal Intent that the said Trustees their Heirs and Assigns and the Survivors and Survivor of them, their or his Heirs or Assignes, shall and may before any other Location; Locate and lay out the five Hundred Acres of Land with the Usual Allowance for Highways holden by of thes Presents, in Trust as a Glebe for the Use of the Minister of the Gospel and Schoolmaster as aforesaid, in such part of the said larger Tract of Land hereby granted as to the said Trustees their Heirs or Assigns or the Survivors or Survivor of them his or their Heirs or Assignes shall seem most convenient and proper to answer the good Ends we thereby propose; any thing herein before contained to the contrary thereof in any wise notwithstanding

PROVIDED also and it is our further Royal Intent and Purpose in this our Grant, that if the Persons their Heirs or Assignes to whose Use the said several and respective Quantities and Lots of Land aforesaid are by Virtue of these Presents held in Trust, and directed to be conveyed as aforesaid, or any or either of them, shall not within one Year after Publick Notice given by the said Trustees their Heirs or Assignes, or the Survivors or the Survivor of them, their or his Heirs or Assignes, in all the News Papers of our said Province, apply to the said Trustees their Heirs or Assignes or the Survivors or Survivor of them his or their Heirs or Assignes, or to some or one of them, for the respective Conveyances above directed to be made to them respectively for their several Shares and Lots aforsaid, and pay or cause to be paid to the said Trustees their Heirs or Assigns, or the Survivors or Survivor of them their or his Heirs or Assignes, or some or one of them, the Charge and Cost of such Conveyances respectively, and their respective proportionable Part of all such Fees and expences as the said Trustees their Heirs or Assigns or the Survivors or Survivor of them, their or his Heirs or Assignes shall be put to, Disburse and expend in

29

and for obtaining this our Grant and in making the Division of the said Lands: With lawfull Interest as aforesaid, that then and in such case it shall and may be lawfull for the said Trustees their Heirs or Assignes or the Survivors or Survivor of them their or his Heirs or Assignes, to sell and dispose of the share and shares Lot and Lots of the said Persons so failing and to convey a good Estate in Fee Simple of in and to the said Share and Shares Lot and Lots respectively to such Person or Persons as shall purchase the same: And out of the Monies arising by the Sale thereof respectively to pay and reimburse themselves the said Proportionable Share of the Costs and Expences aforesaid, with lawfull Interest as aforesaid, and shall render the Overplus Money, if any there be, to the said Person or Persons so failing their Heirs or Assignes, and the Person or Persons their Heirs and Assignes respectively, for whose Use the same is hereby held in Trust as aforesaid, shall be by the said Sale and Conveyance respectively debarred and estopped from claiming or in any wise having any Right or Title in Law or Equity to such the said several Shares and Lots as shall be so sold and conveyed: and the said sales and Conveyances respectively we do hereby consent and agree shall to all Intents and Purposes whatsoever be good against us our Heirs or Successors, any Thing herein before contained to the Contrary thereof in any wise notwithstanding

AND WHEREAS divers of the Persons for whose Use divers Quantities of the Lands hereby granted are respectively held in Trust as aforesaid by these Presents, being Women, who are mentioned and named in these Presents, by their Maiden Names may at the Time of Issuing this our Grant be lawfully married, and thereby their respective Names be changed to the Names of their respective Husbands AND WHEREAS also divers others of the said Persons may now be dead, it is therefore our further Will and Pleasure, and we do hereby direct limit and appoint in every Case where any the said Women have married or are now married that the said Trustees their Heirs or Assigns, or the Survivors or Survivor of them their or his Heirs or Assigns, upon Application and Payment of a Proportionate share of the Fees and Expences Costs and Charges aforesaid, within the Time and in the manner herein before mentioned, with Interest as aforesaid shall convey the respective Lots and Quantities of Land so holden in Trust for them respectively, to them their Heirs and Assignes respectively forever by the Name they shall respectively bear, or be Known by at the Time or Times the same shall be conveyed to them respectively. And in Case any the said Persons shall be dead as aforesaid, then it is our Royal Will and Pleasure and we do hereby direct limit and appoint that the said Trustees their Heirs or Assigns or the Survivors or Survivor of them, their or his Heirs or Assigns shall upon application and Payment of a proportionable Share of the said Costs, Fees Expences and Charges within the Time and in the Manner herein before mentioned

THE ARGYLE DOCUMENTS

with Interest as aforesaid, convey the respective Share and Shares Lot
and Lots hereby holden in Trust for such Person and Persons respectively to the Heirs of such Persons so dead, and to their Heirs and
Assigns for ever.

AND the said several Conveyances and each of them respectively shall
be good and effectual in the Law to all Intents, constructions and purposes whatsoever against us our Heirs and Successors and all and
every other Person and Persons whatsoever claiming or to claim the
same or any Part thereof by Virtue of these Presents, any thing herein
before contained to the contrary thereof in any wise notwithstanding
AND WE do of our especial Grace certain Knowledge and meer Motion, create, erect and constitute the large Tract or Parcel of Land
hereby Granted, and abutted bounded and described in Manner and
Form as is herein before particularly set forth, and every Part and
Parcel thereof, a TOWNSHIP for ever hereafter to, be, continue and
remain, and by the Name of ARGYLE forever hereafter to be called and
Known

AND for the better and more easily carrying on and managing the
Publick Affairs and Business of the said Township our Royal Will
and Pleasure is, and WE do hereby for us our Heirs and Successors
give and grant to the Inhabitants of the said Township all the Powers
Authorities Privileges and Advantages heretofore given and granted
to, or legally enjoyed by all, any, or either of our other Townships
within our said Province; and We also ordain and establish that there
shall be forever hereafter in the said Township, one Supervisor, Two
Assessors, one Treasurer, Two Overseers of the Highways Two Over
Seers of the Poor, one Collector, and Six Constables, elected and chosen
out of the Inhabitants of the said Township yearly and every Year on
the first Tuesday in May, at the most Publick Place in the said Township, by the Majority of the Freeholders thereof, then and there met
and assembled for that Purpose Hereby declaring that wheresoever the
first Election in the said Township shall be held, the future Elections
shall forever thereafter be held in the same Place as near as may be,
and giving and granting to the said Officers so chosen Power and
Authority to exercise their said several and respective Offices, during
one whole Year from such Election, and until others are legally chosen
and Elected in their Room and Stead, as fully and amply as any the
like officers have or legally may use or exercise their Offices in our
said Province, and in Case any or either of the said Officers shall die
or remove from the said Township, before the Time of their Annual
Service respectively shall be expired, or refuse to act in the Offices for
which they shall respectively be chosen: Then our Royal Will and
Pleasure further is, and we do hereby direct ordain and require the
Freeholders of the said Township to meet at the Place where the
annual Ellection shall be held for the said Township, and chuse other
or others of the said Inhabitants in the place and Stead of him or

31

them so dying Removing or refusing to Act within forty Days next after such Contengency.

AND to prevent any undue Election in this Case: WE do hereby ordain and require that upon every Vacancy in the Office of Supervisor; the Assessors, and in either of the other Offices, the Supervisor of the said Township shall within ten Days next after such Vacancy shall happen, appoint the Day for such Election, and give Publick Notice thereof in writing under his or their Hands by affixing such Notice on the Church Door, or other most Publick Place in the said Township, at the least Ten days before the Day appointed for such Election.

AND in Default thereof we do hereby require the Officer or Officers of the said Township or the Survivor of them, who in the order they are herein before mentioned shall next succeed him or them so making Default, within ten Days next after such Default to appoint the day for such Election and give Notice thereof as aforesaid Hereby giving and granting that such Person or Persons as shall so be chosen by the Majority of such of the Freeholders of the said Township as shall meet in the manner hereby directed, shall have, hold, Exercise and Enjoy the Office or Offices to which he or they shall be so elected and chosen, from the Time of such Election, until the first Tuesday in May then next following, and until other or others, be legally chosen in his or their Place and Stead, as fully as the Person or Persons in whose Place he or they shall be chosen, might or could have done by Virtue of these Presents; AND WE Do hereby Will and direct that this Method shall for ever hereafter be used for the filling up all Vacancies that shall happen in any or either of the said Offices between the annual Elections above directed.

PROVIDED ALWAYS and upon Condition, nevertheless that if the said Trustees or the Persons for whom the said several Quantities and Lots of Land are by Virtue of these Presents held in Trust as aforesaid, or some or one of them, their or some or one of their Heirs or Assignes, shall not within four Years next after the Date hereof Settle on the said larger Tract of Land hereby granted so many Families as shall amount to one Family for every Thousand Acres thereof, or if they the said Trustees or the Persons for whom the said several Quantities and Lots of Land are hereby held in Trust as aforesaid or some or one of them, their or some or one of their Heirs or Assignes shall not also within four Years to be computed as aforesaid Plant and effectually Cultivate at the least three Acres for every fifty Acres of such of the hereby granted Lands as are capable of Cultivation, or if they the said Trustees, or the Persons for whom the said several Quantities and Lots of Land are held in Trust as aforesaid or any of them, their or any of their Heirs or Assignes or any other Person or Persons by their or any of their privity consent or Procurement shall fell cut down or otherwise destroy any of the Pine Trees by these Presents reserved to

THE ARGYLE DOCUMENTS

us our Heirs and Successors, or hereby Intended so to be, without the
Royal Licence of us our Heirs or Successors for so doing first had and
obtained, that then and in any of these Cases this our Present Grant
and every thing therein contained shall *shall* cease and be absolutely
void, and the Lands and Premises hereby granted shall revert to and
Vest in us our Heirs and Successors, as if this our present Grant had
not been made: Any thing herein before contained to the contrary in
anywise notwithstanding.

PROVIDED Further and upon Condition also Nevertheless, and we do
hereby for us our Heirs and Successors direct and appoint, that this
our present Grant shall be registered and entered on Record within six
Months from the Date hereof in our Secretary's Office in our City of
New York, in our said Province in one of the Books of Patents there
remaining; and that a Doquet thereof shall be also entered in our
Auditors Office there for our said Province, and that in Default thereof
this our Present Grant shall be void and of none Effect, anything
before in these Presents contained to the contrary thereof in any wise
Notwithstanding. And WE do moreover of our especial Grace certain
Knowledge and meer Motion consent and agree that this our Present
Grant being registered recorded and a Docquet thereof made as before
directed and appointed shall be good and effectual in the Law to all
Intents Constructions and Purposes whatsoever against us our Heirs
and Successors, Notwithstanding any Misreciting, Misbounding, Mis-
naming or other Imperfection or Omission of in or in anywise concerning the above granted or hereby mentioned or intended to be
granted Lands, Tenements, Hereditaments and Premises or any Part
thereof, and of in or in anywise concerning the Trustees aforesaid or
the Persons for whose Use any the said Lands and Premises are held
in Trust as aforesaid or any of them.

IN TESTIMONY whereof we have caused these our Letters to be made
Patent, and the Great Seal of our said Province to be hereunto affixed.

WITNESS our said Trusty and Well beloved Cadwallader Colden
Esquire our Lieutenant Governor and Commander in Chief in and
over our said Province of New York and the Territories depending
thereon in America: at our Fort in our City of New York the thirteenth
Day of March in the Year of our Lord One thousand seven Hundred
and sixty four and of our Reign the fourth. First Skin Line 36 the
Word *last* interlined. Second Skin Line 47 the Word *from* wrote on
an Erazure. 3d. Skin Line 1st. the Word *and* Wrote on Erazure
and Line 6, the Words *and Businesss* and the word *us* and Line 22 the
Words *shall be chosen* interlined.   Clarke

In the preceeding Certificate and Letters Patent recorded for the
Trustees therein named, the following Interlineations &ca. appear—
Page 2 Line 3 the Word *the*, Line 36 (then along the said East Bounds
of the said Tract surveyed for Donald Campbell and others); Page 4

33

APPENDIX

Line 1 the Word *his*: Line 12 the Word *the*: Page 5 Line 20 the Word *said*; and Line 21 the Word *the* interlined; Page 8 Line 30 the Word *and* wrote on an Erazure Page 9 Line 35 the Word *said* Page 11 Line 5 the Word *the* Line 19 Word *or* Line 20 the Word *so* and Line 35 the Word *the* Page 12 line 10 the Word *other*: line 17 the Word *contrary*; and Line 38 the Word *Assigns* Page 13 Line 7 the Words (or Assigns and the Person or Persons their Heirs) Page 15 Line 1 the Word *Days*; and Page 16 Line 19 the Word *a* interlined; and Line 36 *Word* wrote on an Erazure— Examined with the Original this 21th. Day of March 1764 By   G w Banyar   D Secry.

Document VII

GRANTEES NAMED IN THE ARGYLE PATENT WITH THEIR HOLDINGS

LOT	NAME	ACRES	LOT	NAME	ACRES
77	Alex. Montgomery	600	71	Archibald McNiven	350
32	Alex. McNaughton	600	86	John McArthur	350
87	Neil Shaw	600	34	William Fraser	350
22	Dougall McCallor	550	124	William Clark	350
108	James Gillis	500	15	Ann McDuffie	350
107	Neil McDonald	500	..	Arch. McCollum Sr.	350
59	John McEuen	500	31	Donald McIntire	350
75	Duncan McDougall	500	76	Alexander Chrstie	350
109	Archibald McDougall	450	132	Daniel Shaw	350
64	Neil McEachron	450	6	Duncan McKinvan	350
127	Donald McMillan	450	51	Charles McArthur	350
70	John Reid	450	28	Alexander Campbell	350
102	John McArthur	400	43	Elizabeth McNeil	300
95	John McDougall	400	130	John Shaw Ju.	300
69	Hannah McEuen	400	12	Dougall McAlpine	300
119	Dougall Thompson	400	37	Neil McFaden	300
23	Daniel Johnston	350	126	Mary Anderson Sr.	300
111	Patrick McArthur	350	125	Barbara McAllister	300
5	Mary Campbell	350	3	Allan McDonald	300
122	Duncan Taylor	600	17	Archibald McGowne	300
20	Duncan Read	600	48	John McFail	300
104	Malcolmn McDuffie	550	25	William Hunter	300
82	Edward McCallor	500	52	Duncan McFadden	300
138	Duncan Gilchrist	500	65	Hugh Montgomery	300
81	Duncan McGuire	500	120	Mary Graham	300
44	Duncan McArthur	450	19	Duncan McDuffie	350
40	Robert Campbell Jr.	450	131	Duncan Lindsay	350
4	Neil Gillaspie	450	..	Arch. McCollum Jr.	350
36	Duncan Campbell Sr.	450	135	John McIntyre	350
16	Ranald McDougall	400	29	Daniel Clark	350
101	George McKinzie	400	92	Dougall McDuffie	350
42	John McGuire	400	61	James McDonald	350

Maria Campbell 79 300	Alexander Montgomery 78 250	Alexander Montgomery 77 600	Alexander Christie 76 350	Duncan Mc Dougall 75 500	John Cargill 74 300			James Cargill × 73 300	Rachel Nixon × 72 300
Angus Mc Dougall × 80 300	Duncan Mc Glenn 81	Edward Mc Caller 82 500	Hew Gilchrist 83 300	Arch Mc Collum 84 350	Archibald Mc Cort 85 300		John Mc 60 300		Mary Bell
James Nutt × 91 300	Elizabeth Roy 98 300	Roger Mc Neil 99 300	Duncan Campbell 90 300	Neil Shaw 87 600	John Mc Carter 86 350			59
Dougall Mc Duffie 92 350	George Campbell 93 300	James Andrew 94	John Mc Dougall 95 400	Anna Car × 96 300	Charles Mc Allister × 97 300		Florence Mc Kenzie 47	John Mc 48 300
Margaret Mc Neil 103	John Mc Carter 102 400	George Mc Kenzie 101 400	James Campbell 100 300	Hugh Mc Dougall × 99 300	William Graham × 98 500			John Jerry + 45 300
Malcom Mc Duffie 104 550	Florence Mc Varies × 105 300	Arch Mc Ewen × 106 300	Neal Mc Donald 107 500	James Gillis 108 500	Archibald Mc Dougall 109 450		27 × 200		50
Edw Mc Coy 115 300	Angus Graham × 114 300	John Shaw Jen 113 300	John Mc Gowen 112 300	Patrick Mc Arthur 111 350	Marion Mc Bum 110 200		Duncan Campbell Jr × 26 300	William Hunter × 25 300
Duncan Campbell Jun × 116 300	Jennet Ferguson 117 350	Kath Mc Cleary 118 300	Dougall Thompson 119 400	Mary Graham × 120 200	Robert Mc Alpin 300 21		Arch Mc Gowen + 17 300	Eleanor Thompson + 16 300
Donald Mc Mullen 127	Barbara Mc Allister + 125 300	William Clarke 124 350	Elizabeth Calwell 123 250	Duncan Taylor 122 600			Donald Mc Dougall 16 400	Ann Duff 15
Duncan Shaw + 128 300	Alexander Mc Dougall 129 350	Mary Anderson Jon. 126 450	John Shaw Jun 130 300	131 300	Daniel Shaw 132 350	John Campbell + 133 300	Ann Mc Arthur 7 300	8 300
John Mc Arthur 134 300	John Mc Intire 135 350						Duncan Mc Kinnon 6 350	
Mary Ham mells 137 250	Cath Mc 136 250							

Lands Granted & others. Moses Kill

Hudson's River.

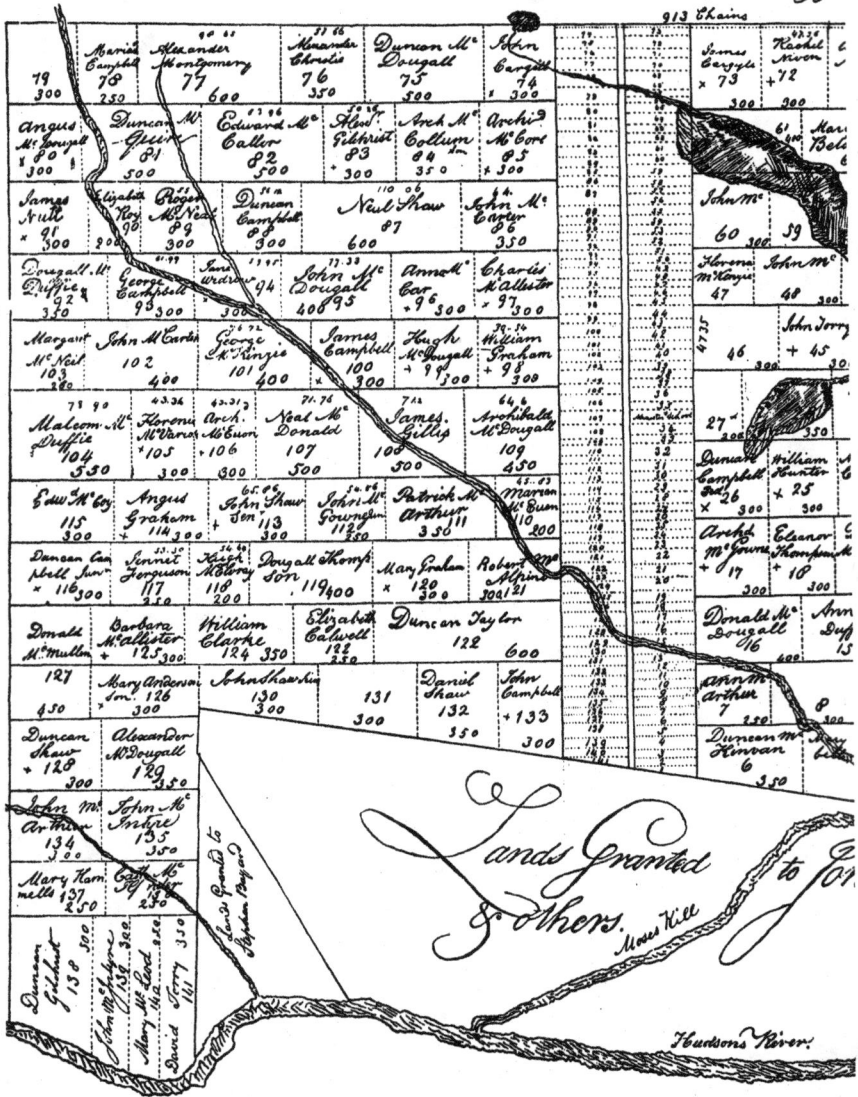

Outline Map of Argyle Patent, with names of the Lot Owners, now published from the original survey, made in 1764 by Archibald Can... and Christopher Yates.—"The Fort Edward Book," by Robert O... com, 1903. (Keating, Pub., Fort Edward, N. Y.)

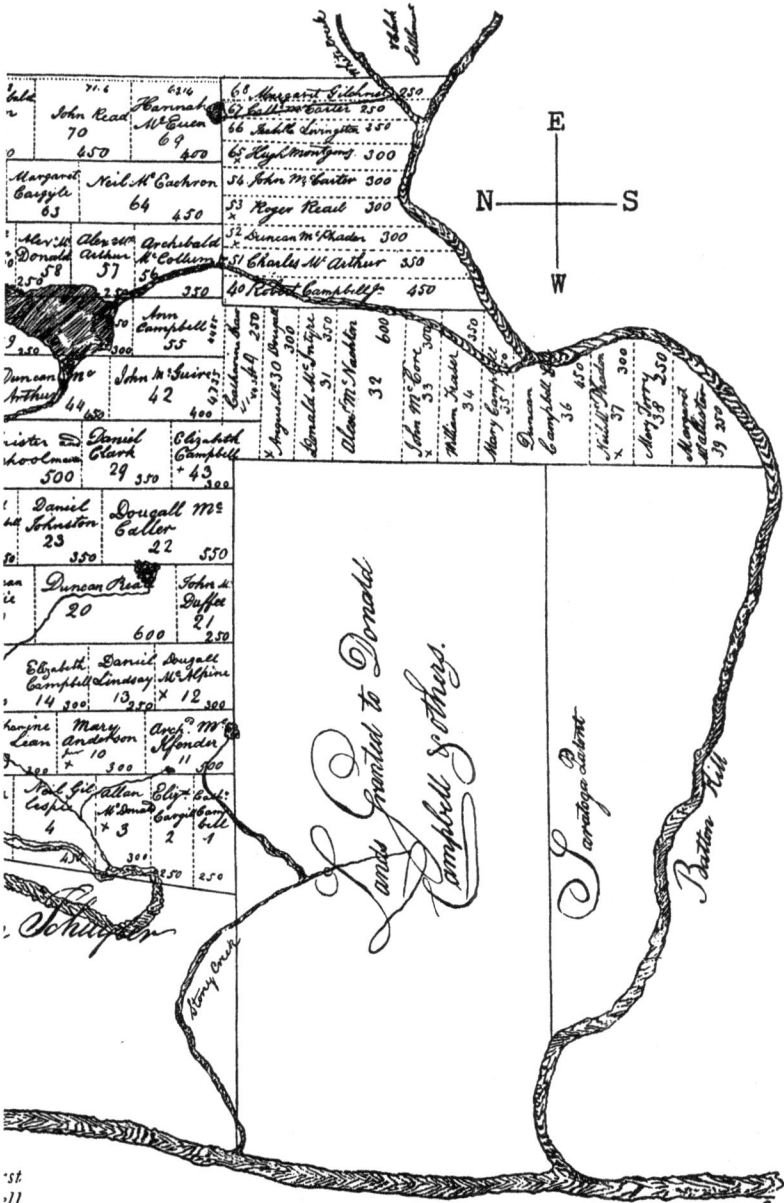

John Read 70 — 650 | Hannah McEuen 69 — 400 | 71.6 | 62.14

68 Margaret Gilchrist 250
67 Sally McCarter 250
66 Archibald Livingston 250
65 Hugh Montgomery 300
56 John McCarter 300
53 Roger Read 300
52 Duncan McPhader 300
51 Charles McArthur 350
40 Peter Campbell 450

Margaret Cargyle 63 | Neil McCachron 64 — 450

Alex McDonald 58 — 250 | Alex Arthur 57 | Archibald McCollum 56 — 350 | 2.50

Ann Campbell 55

Duncan McArthur 44 — 650 | John McGuire 42 — 400

... Margaret ... 30 | Donald McTaggart 31 | Alex McMartin 32 — 600 | John McIver 33 | ... 34 | Mary ... 35 | Duncan Campbell 36 | Neil McMartin 37 — 300 | Mary ... 38 — 350 | Margaret ... 39 — 250 | 460

schoolman 500 | Daniel Clark 29 — 350 | Elizabeth Campbell 43 — 300

Daniel Johnston 23 — 350 | Dougall McCaller 22 — 550

Duncan Read 20 — 600 | John McDuffee 21 — 250

Elizabeth Campbell 14 — 300 | Daniel Lindsay 13 — 250 | Dougall McAlpine 12 — 350

McLean | Mary Anderson 10 — 300 | Archd McFonder 11 — 300

Neil Gillespie 4 | Allan McDonald 3 — 300 | Elizabeth Cargil 2 — 250 | Saml Cargilbaum 1 — 250

Schuyler

Stony Creek

Lands Granted to Donald Campbell & others.

Saratoga Patent

Button Hill

E — N — S — W

THE ARGYLE DOCUMENTS

LOT	NAME	ACRES
129	Alexander McDougall	350
..	Archibald Campbell	300
115	Edward McCoy	300
91	James Nutt	300
10	Mary Anderson Jr	300
97	Charles McAlister	300
100	James Campbell	300
113	John Shaw Sr	300
..	Neil Carmicheal	300
141	David Torry	300
134	John McArthur	300
..	John McCallman	300
62	Mary Beaton	300
98	William Graham	300
50	John McGours Sr	
99	Hugh McDougall	300
72	Rachel McNiven	300
83	Alexander Gilchrist	300
106	Archibald McEuen	300
11	Archibald McIlfender	300
80	Angus McDougall	300
121	Robert McAlpine	300
88	Duncan Campbell (3)	300
55	Ann Campbell	300
73	James Cargill	300
94	Jane Widrow	300
114	Angus Graham	300
96	Ann McArthur	300
33	John McCore	300
93	George Campbell	300
18	Eleanor Thompson	300
13	Daniel Lindsay	250
140	Mary McLeod	250
123	Elizabeth Caldwell	250
66	Isabella Livingston	250
24	Arch. Campbell Jr	250
57	Alexander McArthur	250
30	Angus McDougall	300
79	John Gilchrist	300
9	Catharine Mclean	300
105	Florence McVarich	300
53	Roger Reed	300

LOT	NAME	ACRES
46	Malcolm Campbell	300
116	Duncan Campbell Jr	300
14	Elizabeth Campbell	300
139	John McIntyre	300
74	John Cargill	300
133	John Campbell	300
89	Roger McNeil	300
45	John Torry	300
85	Archibald McCore	300
128	Duncan Shaw	300
..	Hugh McCarty (McArthur)	300
103	Margaret Mc Neil	250
21	John McDuffie	250
117	Janet Ferguson	250
112	John McGowne Jr	250
78	Marion Campbell	250
58	Alexander McDonald	250
1	Catharine Campbell	250
41	Catharine Shaw	250
38	Mary Torry	250
2	Elizabeth Cargill	250
7	Ann McArthur Sr,	250
39	Margaret McAlister	250
47	Florence McKinzie	200
118	Hugh McIlvray	200
27	Elizabeth Fraser	200
8	Mary McGowne	300
54	John McArthur	300
49	Jane Cargill	250
136	Catharine McIlfender	250
35	Mary Campbell	250
63	Margaret Cargill	250
137	Mary Hammell	250
68	Margaret Gilchrist	250
67	Catharine McArthur	200
110	Marion McEuen	200
90	Elizabeth Roy	200
26	Duncan Campbell Jr	300
60	John McMitchell or John McEacron	300

Four grantees named in the Argyle patent, do not appear in the printed lists of the grantees of that patent, namely Neil Carmicheal, Archibald McCollman, Hugh McCarty or McArthur, and Archibald Campbell Sr.

35

Document VIII

(Endorsement) A List of the Persons Brought from Scotland by Captain Lauchlin Campbell to settle the Kings Lands at the Wood Creek from 1738 to 1740—89 Familys.............358
Persons 112 Single 112
                                            _____
                                            470 Persons

Memorandum of Passengers who Came in the years 1738, 1739 & 1740

A List of Passengers from Islay with Captain Lauchlin Campbell bound for New York, July 1738.

No Claim Ronald Campbell, Dead. John Campbell of Balinabie & Anna Campbell his wife. Alexander Montgomery & Anna Sutherland his wife. Hugh Montgomery. Mary Beaton. Duncan McEuen. Janet McEuen, (son & Daughter to Hugh McEuen). Mary McEuen. Mary McEuen, Daughter to John McEuen. Janet Ferguson (her son Alexander McDonald). Archibald Johnston & Christine Johnston his wife. No Claim Mary Graham, Dead. John McNeil & Eliz: Campbell his Wife, & Barbra, Peggie, Catharine, Betty & Neil, 5 Children. Margaret McNeil. Angus McAlister. Elisbie Thompson of Dunardrie. No Claim Alexander McLean, Died at Cuba. No claim William Campbell, Dead family But in Scotland. No Claim William Campbell Wheel Wright, Dead. Alexander Graham. Donald Carmichael & Elizabeth McAlister his wife. John, Alexander & Mary his three Children. James Campbell & Anna McDougall his wife. Archibald Lauchlin Eliz; & Janet his 4 Children. Neil McArthur & Mary Campbell his Wife & Alexander. John and Christian his 3 Children. Donald Shaw & Merran McInish his Wife. Mary Campbell. Elisbie Sutherland and her Children, James, Alexander, Duncan, Margaret & Elizabeth Gillies 5 Children. Duncan Taylor & Mary Gillies his Wife & Mary his Daughter. Archibald McEchern & Jean McDonald his wife & Catharine his Daughter. Donald McMillan & Mary McEachern his Wife. Donald McCloud & Catharine Graham his Wife, John & Duncan his 2 sons. Ronald McDougall & Bettie McDougall his Wife, John & Alexander his 2 sons. Allan McDougall & Elizabeth Graham his Wife, Margaret, Anna & hannah his 3 Daughters. Archibald McDougall & Christian McIntyre his Wife, Alexander & John his 2 sons. Hugh McDougall. Archibald McKellar & Jannet Reed his Wife. Charles McKellar & Florence McEachern his Wife, Margaret, Catharine & Mary his 3 Daughters. Catharine Fraser. Alexander McNaught(on) and Mary McDonald his Wife, John, Moses, Janet & Eleanor his 4 children. John McNiven & Mary McArthur his Wife, Elizabeth & Mary his 2 Daughters. Merran McNiven. Rachel McNiven. Patrick McArthur & And Mary McDougall his Wife, Charles, Colin & Janet his 3 Children. Duncan McArthur &

36

THE ARGYLE DOCUMENTS

Anna McQuin his Wife, Anna, Mary, Margaret & John 4 Children.
Alexander McArthur & Catharine McArthur his Wife, John, Donald,
Duncan, Catharine & Florence his 5 Children. Donald McEachern
and Anna McDonald his Wife, & Catharine his Daughter. Neil Mc-
Donald & Anna McDuffie his Wife, Donald, Archibald & Catharine
his 3 Children. Duncan Gilchrist & Florence McAlister his Wife &
Mary his Daughter. John McKenzie & Mary McVurrich his Wife,
Archibald & Florence 2 Children. George McKenzie & Catharine Mc-
Niven, his Wife, Donald & Colin his 2 sons. Malcolm McDuffie &
Rose Docharty his Wife, Margaret & Janet his 2 Daughters. Dudley
McDuffie & Margaret Campbell his Wife & Archibald a son. John
McIntagart. Malcolm Martine & Florence Anderson his Wife. Dugald
McAlpine & Mary McPhaden his Wife, Donald & Mary his 2 Chil-
dren. John McIntaylor. James Stewart. Donald Campbell & Mary
McKay his Wife, Robert, James, Margaret & Isabel his 4 Children.
William McGie. Duncan Smith. James Livingston. John Gilchrist.
Alexander Gilchrist. Lauchlin McVuirich. Alexander Campbell.
Allen Thompson. Donald McIntyre. Murdoch Hammel. Donald
McIntaylor. John McColl. John McLean. Christain Paterson. Cath-
arine Lessly. Mary Ross. Jean Ross. Merran Hameel.
33 Familys, 42 passengers, 177 Persons.

Passengers from Islay, June 1739.

Robert Fraser & Mary McLean his Wife, Charles, Coline, Sarah,
Catharine, Mary & Isabel 6 Children. Archibald McEuen & Janet
McDougall his Wife. Malcom McEuen. James Nutt & Rebecca
Creighton his Wife, Robert, John & Elizabeth his 3 Children. Neil
Campbell. Peter Green. John Caldwell & Mary Nutt his Wife,
Alexander & James his 2 Sons. Neil McPhaden & Mary McDearmid
his Wife, Dirvorgill & Margaret his 2 Daughters. Angus McIntosh.
Alexander McChristen. Catharine Campbell. Jean Cargill. Florence
McVurich. Archibald McVurich & Merran Shaw his Wife. Neil
Shaw. Catharine Shaw. John McQuary & Anna Quarry his Wife.
Patric McEachern & Mary McQuarry his Wife. Donald McPhaden.
Dugald Thomson & Margaret McDuffie his Wife, Archibald, Duncan
& Christie & his Brothers Daughter 4 Children. Patrick Anderson &
Catharine McLean his Wife. Duncan Campbell & Sarah Fraser his
Wife. Charles McAlister & Catharine McInnish his Wife, John &
Margaret his 2 Children. Duncan McAlister & Effie Keith his Wife.
Donald Ferguson & Flory Shaw his Wife With One Child of his Own
& Catharine & Anna Ferguson his Brothers Children. William Clark,
his Wife & one son John. Donald Livingston & Isabel McCuarg his
Wife, John & Duncan his 2 Children. John McEuen & Anna John-
ston his Wife & his son Malcolm. Lauchlin McVurich. John Mc-
Donald. James Cameron. Mary Thompson. Murdock McInnish &
Merran McKay his Wife, Catharine, Archibald, Neil, Anna & Flor-

37

ence his 5 Children. Archibald McDuffie & Catharine Campbell his Wife, John and Duncan his 2 sons. Neil McInnish & Catharine McDonald his Wife. Duncan Reid & Mary Semple his Wife, Alexander Nicholas, Angus & Jennie his 4 Children. Neil Shaw & Florence McLachlin his Wife. John Shaw & Mary McNeil his Wife, Neil & Duncan his 2 sons. Gustavus Shaw. Archibald McGown with his 3 Children, Duncan, John & Margaret. Malcolm McGown with Patrick alias Hector his 2 Children. John McGown & Anna McCuarg his Wife, Malcolm and Angus his 2 sons. Donald McMillan & Janet Gillies his Wife & Alexander his son. Alexander McDuffies Widow, Anna Campbell (he Dieing at sea) Archibald, Duncan, James, Mary & Isabel his 5 Children. Duncan McQuarrie alias Brown & Effie McIlepheder his Wife, Donald, John, Gilbert & Christian his 4 Children. Archibald McIlepheder. Catharine McIlpheder. Donald Lindsey & Mary McQuarrie his Wife, Richard Duncan, Effie & Christion his 4 Children. Neil Gillaspie & Mary McIlepheder his Wife, Gilbert and Angus his 2 Sons. John Reid & Margaret Hyman his Wife and his son Donald. Roger Reid. Dugald Carmicheal & Catharine McEuen his Wife, Janet, Mary, Neil & Catharine his 4 Children. Merran McEuen with her Daughter. Christain McAulla. Patrick Robertson. Duncan McDougall & Janet Calder, his Wife, John Alexander, Ronald, Dugald & Margaret his 5 Children. Dugald Gilbert, Flory & Margaret his 3 Children. Archibald McCollum & Merran McLean his Wife, Donald, John, Margaret, Mary & Allan his 5 Children. James Torry & Florence McKay his Wife & his Children Mary & Catharine. Nicholas McIntyre & Margaret Peterson his Wife & John his son & Catharine. George Torry. Cornelius Collins. Angus McDougall. Alexander Hunter & Anna Anderson his Wife, his Children, William, Alexander & Janet. Alexander McArthur & Catharine Gillies his Wife, Duncan & Flora his 2 Children. Angus Campbell with his son John. John McPhail & Christy Clark his Wife, McIntyre.

42 families, 24 Single Passengers, 193 Persons.

Passengers from Islay, November 1740.

Neil Campbell. Edward Graham & Jean Fraser his Wife. John McEuen. William Adair. Malcolm Campbell. Alexander Campbell & Margaret Campbell his Wife & One Daughter Merran. Duncan Campbell of ye family of Duntroon. Alexander Campbell of ye family of Landie. Duncan Campbell & anna Campbell his Wife (Lenos) and one Daughter Catarine. Robert McAlpine. Duncan Campbell of Lochnel. William Campbell, Archibald Campbell, of Ardenton. Anna Campbell. Duncan Campbell of ye family of Dunn. Duncan McCollum. John McIntyre. John Christy & Isabel McArthur his Wife, Hannah & Mary his 2 Daughters. John McArthur & his son Neil, Daughter Christian. Angus Clark & Mary McCollum his

Wife & Catharine & Mary his 2 Daughters. Anne McNeil Widow
to Hugh McEuen, with her Son Alexander and Mary her 2 Children.
Elizabeth Cargill. James Cargill. John Cargill. David Cargill.
Margaret Cargill. Ann McArthur. Jean Widrow. Merran Mc-
Indeora. James McEuen. John Shaw & Merran (Sarah) Brown his
Wife. Donald Mary & one Infant (Margaret born at sea). Chris-
tian Brown. John McGibbon. Archibald Graham. Roger Thomp-
son. John Campbell. Duncan McKinven & marian McCollum his
Wife & Donald & Mary his 2 Children. John McGilvrey & Catarine
McDonald his Wife, Hugh Donald, Bridget & Mary his 4 Children.
Anthony Murphy. Duncan McKay. Dudley McDuffie & Margaret
McDougall his Wife, Dugald & Mary his 2 Children. Duncan Mc-
Phadden & Flory McCollum his Wife, John and Duncan his 2 Chil-
dren. Archibald McCollum & Flory McEacheon his Wife, Hugh &
Duncan his 2 Sons. Archibald Hammel. Mary Hammel. Catharine
Graham. Margaret McArthur Wife to Archibald McCollum at New
York & Anna & Mary his 2 Daughters. Mary Anderson, Widow with
her 2 Children. Duncan Leech & Mary Leech. Margaret McAlister.
Effie McIlvrey. Lauchlin McLean. Angus Graham. Roger McNeil.
John Reid. Ann McArthur.

Document IX

*List of Persons brought from Scotland by Capt. Laughlin Campbell
in 1738-40. This list was probably prepared in 1763.*

Heads of Families Imported in 1738.

1 Ronald Campbell Decd. No family with him.
2 John Campbell Dead, brought a Wife with him who is
Dead and they have left no Children. but he has a sister
called Ann, who is married in the Highlands & has 5 Childn.
married Duncan Campbell who is in this list hereafter.
3 Alexander Montgomery, now living, has a Wife and no
Children 200
4 Archibald Johnston he is Dead His Wife Kerstain Johnston is living as also two Sons and three Daughters. she is
married to Daniel Mc Alpine. Macolm Johnston for himself one Bror. & three Sisters...................... 250
5 John McNeil he is Dead, his Widow is living and four
Daughters in this Province and one in England. N:B: one
of the Daughters Named Jane came over in 1740 four
Daughters. 200
6 Donald Caemichael, he is Dead, has Children, but hath none
in this Province.
7 James Campbell He is Dead, His Widow Anna McDougall
and one Son Archibald and two Daughters are living. Widow
(100), Archibald the son (50), Isabel (50)............. 200

8 Neil McArthur Dead, His Widow and five Children Living 300
9 Donald Shaw Dead Son and Daughter living........... 200
10 Elizabeth Sutherland She is living and four Children..... 400
11 Duncan Taylor living, and has a Wife and Eight Children three of them married............................. 500
12 Archibald McEachern brought a Family Consisting of a Wife and Daughter.
13 Donald McMillan living, has a Wife and five Children, two of whom are Married.......................... 400
14 Donald McCloud is Dead, & has one Daughter living..... 150
15 Cormick McCoy Dead his Widow living and a Son and Daughter who are Married........................ 200
16 Ronald McDougall living, with a Wife & two Children who are both Marriwd. John one of his Sons is Dead & hath left two Children............................. 300
17 Allan McDougall Dead the Widow, one son and four Daughters living. 300
18 Archibald McDougall, Living, has a Wife and five Children three of whom are Married.................... 350
19 Archibald McKellar Dead hath left a Widow & 8 Children. one Married................................ 450
20 Chas. McKellar Dead hath left a Widow and Seven Children. One of whom is Married. The Mother of these two McKellars Came over with them but is Dead............ 400
21 Alexander McNaught(on) Living, has a Wife and four Children three of whom are Married. he has 8 Grand Children. 500
22 John McNiven Dead One Son and four Daughters living three of whom are Married...................... 250
23 Malcolm Martine Dead his Wife alive not Known
24 Patrick McArthur living, has a Wife two sons and one Daughter. 250
25 Duncan McArthur Dead, his Widow and two sons and two Daughters living, three of whom are Married........... 250
26 Alexander McArthur Dead, two sons and four Daughters four are Married. John the Eldest son Dead and has left a Widow and two Children........................ 350
27 Donald McEchern Dead his Widow and three sons and three Daughters living, two Married.................. 350
28 Neil McDonald living, has a Wife and Six Children. one Married. 400
29 Duncan Gilchrist himself, Wife, and Six Children living, one Married. 400
30 John McKinzie Dead. One Daughter left.............. 100
31 Cormick McCoy before (see No. 15)

40

32 George McKenzie, living with his Wife and four Children, two of them Married. lives in New Jersy 300
33 Malcolm McDuffie, & Wife, living with three sons and four Daughters one Married 450
34 Dudly McDuffie Dead Married and (had) a son & a Daughter the Daughter Married 150
35 Dugald McAlpine and Wife and two Children, who are both Married. 200
36 Donald Campbell Dead, his Wife and four Children living. One of whom is Married 250

Heads of Families Imported in 1739.

1 Robert Fraser Dead, three Daughters living, and William Fraser the son of Charles who was the Eldest son of Robert & two Sisters.. 250
2 Archibald McEuen, Dead, two Children living a Son & Daughter 150
3 James Nutt living, and one son who is Married......... 200
4 John Caldwell doubtfull whether living or not, but has Daughter living in N. York, who is Married, & two sons which he took to Pensilvania.
5 Neil McPhadon and Mary his Wife and one Daughter Called Margaret who is Married & hath 2 Children are now alive. 200
6 Archibald McVarick Dead, his Widow living, who hath two Children living by another Marriage.
7 John Quary and his Wife and four Children, one Married. 300
8 Patrick McEachern Dead, his Widow living........... 100
9 Dugald Tomson and his Wife, three sons, two sons Married. 300
10 Patrick Anderson, Dead, his Widow & two Daughters by him, &c Many Children by another Marriage.......... 200
11 Duncan Campbell, Dead, his Widow & three sons and a Daughter living, the Daughter Married............... 250
12 Charles McAlister Dead, left two sons, three Children of the Eldest son living, & the youngest son............... 200
13 Duncan McAlister, Dead, One son and two Daughters living. 200
14 Donald Ferguson, Dead, one Daughter living, and a Daughter of his Brother whom he brought over............... 150
15 William Clark his Wife and two Children, a Son and Daughter. 250
16 Donald Livingston Dead, his Widow and Daughter Living. 150
17 John McEwen living, and his Wife and five Sons........ 400
18 Murdock McInnish Dead, three Grandchildren by the Widow of his son Neil by another Marriage, and three by his Daughter Florence........................... 200
19 Archibald McDuffie Dead, One son Duncan & two Children & two Children half Blood........................... 150

41

APPENDIX

20 Neil McInnish   the son of Murdock above Mentioned
    Widow Married to Allen McDonald.................... 100
21 Duncan Reid Living brot. over his Wife and 8 Children
    all Dead 500
22 Neil Shaw Dead. Five Grand Children living Neil the
    Eldest   to Youngest............................... 200
    Neil the Eldest.................................... 200
23 John Shaw Dead, Neil and two other Children living two
    Married. provided for above
24 Archibald McGown Dead   A Grandson living Named
    Archibald, and a Daughter who hath 4 sons............. 200
25 Malcolm McGown, hath one son living who is Married &
    hath Children. 150
26 John McGown, & Wife both alive..................... 200
27 Donald McMillan, alive, five Children 3. Sons & 2 Daugh-
    ters 350
28 Alexander McDuffie, Died at Sea, his Widow and two
    Daughters, his son Duncan Duffie who is dead hath left one
    Daughter called Anne............................... 250
29 Duncan McQuore & Wife and Five Children four Sons, &
    One Daughter are living, the four Sons are Married...... 400
30 Donald Lindsey living One Son & two Daughters living
    one son & One Daughter Married..................... 250
31 Neil Gillaspie living as also his Wife & one Daughter.... 350
32 John Reid living as also his Wife and five Children, three
    Boys & one Girl, the Daughter Married............... 350
33 Dugall Carmichael Dead   One son living Named John
    brought over a Numerous family..................... 200
34 Merrian McEuen Dead the above named John the son of
    Dugall is her Nephew.
35 Duncan McDougall Alive as also his Wife & six Children,
    three Sons & two (three) Daughters two sons & a Daughter
    Married the Daughter a Widow & 4 Children.......... 400
36 Dugald Campbell Dead. Archibald Campbell of N. York
    his Heir 150
37 John McPhail Dead, his Widow a son and a Daughter
    living. 200
38 Archibald McCollum   Living, with two sons and One
    Daughter, one son & the Daughter Married............. 250
39 Nicholas McIntire Dead his Widow two sons and two
    Daughters living. 250
40 James Torry Dead, two sons and two Daughters living,
    Daughters Married. 200
41 Alexander Hunter Dead, son & Daughter living who are
    both Married & have Children...................... 200

42

THE ARGYLE DOCUMENTS

42 Alexander McArthur   Dead   his Widow & One Son living brought over a large family........................ 200
43 John Campbell & Mary his Mother   Dead,   Archibald the Nephew of John Living   provided for

Heads of Families Imported in 1740

1 Edward Graham   Dead   One Daughter living. provided for
2 Alexander Campbell   Dead,   two Daughters living...... 150
3 Duncan Campbell   living as also his Wife, with three Sons and two Daughters 350
4 John Christy   Dead   his Widow One Son & three Daughters living, two Daughters Married................... 250
5 John McArthur   Dead   a Son and Daughter living. both Married. 150
6 Angus Clark   Dead,   two sons & One Daughter living. Daniel the Eldest son Dead, leaving a son & Daughter.... 250
7 Anne McNeil   said to be living at Basking Ridge in New Jersey, with her Children.
8 John Shaw   Dead, his Widow and four Children living 2 Sons & 2 Daughters. a son and a (two) Daughters Married. 250
9 Merrian McCollum   the Parties know nothing of this Person at present   she having moved to N. York.
10 John Mc Elvrey   Dead   One son living at Amboy....... 100
11 Dudley McDuffie   Dead, his Widow & two sons and two Daughters living.   One Daughter Married............. 250
12 Duncan McPhaden   Dead   One son John the Eldest Dead Leaving 4 Children & One son Duncan living.......... 200
13 Archibald McCollum   living with a son and a Daughter several Grang Children........................... 250
14 Archibald McColeman   Dead   Widow and one son and two Daughters living. 200
15 Mary Anderson   living, with two Dauthers both married. who have sons grown up............................ 200
16 Duncan McKinven   Living   and one son & three Daughters, one Daughter Married   New York.................... 250

Single Persons Imported in 1738

1 Hugh Montgomery living in N. Y.: & is Married and has two Children 200
2 Mary Beatton   living, is a Widow and has a son Married.. 200
3 Duncan McEuen ⎫
4 Jennet McEuen ⎬ living in the Jerseys and are Married
5 Mary McEuen ⎭ there
6 Mary McEuen   lives in or about the same place & is Married there
7 Jennet Ferguson   Dead, One son living in N.Y:.......... 150

· 43

8 Mary Graham Dead, has Children living in the Manor of Livingston 200
9 Margaret McNeil Living in the Highlands.
10 Angus McAlister Said to be living in Carolina.
11 Alexander Graham Died, has left two sons both in N. Y... 200
12 Hugh McDougall lives in Livingston Manor.......... 200
13 *Merran McNiven 200
14 *Rachel McNiven 200
     *both live in New York and have Children
15 James Livingston Dead, his Widow and Children live in Trenton in Jersey.
16 *John Gilchrist 200
17 *Alexander Gilchrist 200
     *both living, and are Married in the Highlands of this Province
18 Alexander Campbell, is Married & has a family in Amboy.
19 Donald McIntire Lives in New York. one son, & a Wife who likewise came over with Capt. Campbell as p List.... 250
20 Murdoch Hammell lives in the Island of Jamaica.
21 John McLean Has a Relation in Town.
22 *Lauchlin McLean ....
23 *Mary Ross married after their arrival................ 200
     *Died leaving one Child named Catharine Who is now in Albany

Single Persons Imported in 1739.

1 Malcolm McEuen is Dead but has left three Children who now live in New York............................ 200
2 Neil Campbell lives in the Island of Jamaica.
3 Catharine Campbell lives in the Highlands............. 150
4 Jane Cargill Married in New York to Mr. Van Vleck Merchant 150
5 Florence McVarick, is Married and has Children. in Livingston Manor 200
6 Catharine Shaw lives in New York is Married and has one Child 150
7 Mary Thomson Married and lives in Pennsylvania.
8 *Archibald McIlpheder 200
9 *Catharine McIlpheder 150
     *both Married and live in the Highlands & have Children.
10 Roger Reid Married and lives in the Highlands, & has 3 Children 200
11 George Torry Dead, has left one Child in N. York...... 150
12 Cornelius Collins lives in the Jerseys.
13 Angus McDougall is Married and lives in the Highlands. 200
14 David Shaw Dead. Widow living in Tappan.......... 150

44

THE ARGYLE DOCUMENTS

Single Persons Imported in 1740.

1 John McEuen   lives in the Province of Pennsylvania a
   Doctor.
2 Malcolm Campbell   lives in New York, a Merchant...... 200
3 Alexander Campbell   Dead   but has one Daughter alive in
   New York. 150
4 Robert McAlpine   lives in New York, has a family of five
   Sons & 2 Daughters................................ 200
5 Duncan Campbell   Married in New York and has sevl.
   Children. 200
6 *William Campbell 200
7 *Archibald Campbell 200
   *both Dead, but have left Children who live in the High-
            lands.
8 Anne Campbell   lives in the Highlands & is Married & has
   a family of 6 Children.
9 John McIntire   a Clergyman in Pennsylvania.......... 200
   for a place of Worship & School house................. 500
10 Elizabeth Cargill   lives in Tappan and is Married there.. 150
11 *James Cargill 200
12 *John Cargill 200
13 *David Cargill 200
   *All living in N. Y. and are Married & have Children.
14 Margaret Cargill   is a Widow, has Children & lives at
   New Rochell. 150
15 Anna McArthur   Married in Albany.................. 150
16 Jane Widrow   Is Married & has a family of 7 Children in
   the Highlands 200
17 James McEuen   Said to live in Boston.
18 Roger Thomson   Dead. his Widow lives in Amboy & his
   Children.
19 John Campbell   Married in New York................ 200
20 Mary Hammel   Dead, but has a Daughter left who lives
   in the Highlands.................................. 150
21 Margaret McAlister   is Married and lives at the Manor of
   Livingston 150
22 Angus Graham   Lives in New York, has two sons & 3
   Daughters
23 Roger McNeil   Living on Long Island................ 200
24 Anne McArthur   Lives in the Highlands & has five Children 200
25 Margaret Gilchrist   Lives in New York.............. 150
26 John Torry   Married and living in N. York............ 200
   (Colonial Manuscripts, vol. 72, p. 170, New York State Library.)

APPENDIX

Document X

A further Account Delivered by Alexander McNaught(on) and Duncan Reid of Persons who did Emigrate with Captain Campbell in 1738. 1739, and 1740, and who have or their Descendants or persons Impowered, lately appeared and Requested a Proportion of the lands Intended to be Granted. . . . This account was delivered on . . . the 10th of May 1763.

George Campbell of the City of New York Merchant came over in pursuance of a Letter written by Captain Campbells orders to him and dated in 1742, Offering him Incouragement Concerning the Lands then promised. . . .

John McCore came over in 1739, he is now Married and lives in the Highlands. . . .

Archibald McCore came over in 1739. is Married and lives in Tappan. . . .

James McNaught(on) dead, came in 1740 but his Brothers son John McNaught(on) who lives in Tappan, prays his proportion and Engages to settle it . . . provided for before

Duncan Campbell came in 1740, and his Brothers Daughter Mary Ann Campbell of the City of New York, prays his proportion & will Engage to settle it. . . .

Angus McAlister came in 1738, is now living in South Carolina and his Sisters Daughter who is Married to Jacob Vandle of N. Y. will enter into any Engagements necessary during his Absence. . . .

Peter Robertson came in 1739, is dead—his Cousin John McDonald of the City of New York Carpenter prays his proportion and Engages to settle it. . . .

Mary Thompson came in 1739, lives in Pensilvania, her Cousin Duncan Reid of N. Y. praus her proportion and Engages to settle it. . . .

Charles McArthur of the City of N. York with his Wife and Family came on Board in 1738, and the Ship being too much Crouded was turned ashore, and as they had sold all their Effects this Obliged them to go to Ireland where he took a Passage and arrived here a fortnight before the ship in which he first Engaged with Captain Campbell. . . .

Donald McMillan came in 1738, he is now dead & Allan McDonald of the City of New York Tavern Keeper his Kinsman Engages to settle his Proportion. . . .

Neil Campbell lives in Jamaica, came in 1739, Alexander Montgomery of Tappan who is Married to his Mother will take a Grant in Trust for him & Enter into the Necessary Engagements.

Ronald Campbell came in 1738, George Campbell of this Province Pedlar, prays his Proportion being his nearest Relation. . . .

46

THE ARGYLE DOCUMENTS

Donald Campbell now living in Jamaica came in 1738 his Cousin
Duncan Campbell of this City appears to act in Trust for him. . . .
Jennet Ferguson is now dead came in 1738 and her son Alexander
McDonald a Rope Maker in New York prays A grant of her pro-
portion & Engages to settle it. . . .
William Campbell Joiner now dead, came in 1738 his Cousin Alex-
ander McNaught(on) in Tappan prays a Grant of his share & En-
gages to settle. . . .
Catharine Graham came in 1740 died in New York and her Broth-
ers son John Graham of New York prays a Right to her share &c. . .
John McDonald came in 1739, is now at sea, his Cousin Allan
McDonald of N. York will act in his Absence. . . .
John Reid came in 1740 is gone to Virginia, his Uncles Son Peter
Reid in Tappan Engages to Act for him. . . .
Duncan McKay came in 1740, went to sea & is dead, his Cousin
Mary McKay of the City of New York Widdow prays a right to his
share which she Engages to settle. . . .
Margaret Gilchrist came in 1740, lives in New York. . . .
Duncan McCollum came in 1740, died here, and Daniel Campbell
of this City his Cousin prays a right to his share which he Engages
to settle. . . .
William Adair came in 1740, dead, his Cousin Duncan Reid prays
a right to his share which he Engages to settle. . . .
John McIntaylor came in 1738 and his Uncles Son Donald Smith
of the City of New York Mariner prays a right to his proportion
which he Engages to settle. . . .
Archibald McEachern and his Wife Jean McDonald and one
Daughter came in 1738, and his Cousin Finlay McEachern is desirous
of Taking their proportion in Trust until they can be found. . . .
Alexander Christy came in 1738, is dean and his Cousin Mary
Christy who is Married to Duncan Campbell of New York prays a
right to his proportion which he will Engage to settle. . . .
William Campbell Wheel wright came in 1738 is dead and his
Cousin Mary Mackey of the City of New York prays a grant of his
proportion which she will Engage to settle. . . .
Donald McIntaylor came in 1738 is dead and his Cousin Alexander
Taylor in Tappan prays a Grant of his Proportion which she will
Engage to settle. . . .
Jane Ross came in 1738, is Dead, has a Daughter living which is
a Minor, and John Torry of N. York prays a grant of her Propor-
tion which he Engages to settle in trust for the Minor. . . .
Donald McIntyre came in 1738, is Dead leaving Malcolm Graham
of N. York Pruke maker his son. . . .
Malcolm McDuffie camd in 1739 is Dead, & and his Kinsman
Duncan Reid of N. York prays his Proportion which he will settle. . .

47

APPENDIX

Roger Thompson came in 1740 died in the Provincial Service has
left a Widdow & one Child in Amboy, who hath appointed Archibald
Gilchrist of N. York to act for them. . . .
Catharine Fraser came in 1739, is Dead leaving one Daughter called
Elizabeth who lives in New York. who hath appointed her Cousin
Robert Campbell of N. York to act for her. . . .
Mary Fraser came in 1739 and is married & lives in New York. . .
Gustavus Shaw came in 1739, is Dead & his Nephew Neal Shaw of
the City of New York Rope maker, in Trust for the Rest of his Heirs
prays his proportion which he Engages to settle. . . .
Catharine Fraser came in 1738 is Dead, and has left two Grand-
daughters, one named Catharine Montgomery & 'tother Catharine
Stevenson who are both Married & live in New York. . . .
Elizabeth Fraser came in 1739 and is Married & lives in New York.
John McLean came in 1738 is Dead and his Cousin Alexander
McLean of the City of Albany prays his share which he Engages to
settle. . . .
Marian Culbreth came in 1739, is Dead, & Duncan Reid her next
heir Prays her share which he Engages to settle. . . .
Alexander Campbell came in 1738, lives in Amboy, hath applyed
and declares his willingness to settle such Proportion as shall be
granted to him. . . .
John Brady came in 1740, had 5 Children, one of whom named
Hugh lives in Amboy and prays his fathers proportion which he
Engages to settle. . . .
Effie McIlevray came in 1740, and lives in New York. . . .
John McDougall came in 1739, died a privateering in the last war,
his Brother Dougal McDougall of new York prays his proportion &c.
(Endorsement)
(New York Colonial Manuscripts, vol. 72, p. 171, in the New
York State Library, Albany, N. Y.)

Document XI

*On August 12th 1771 Sarah, the widow of John Shaw presented
the following.*

PETITION OF SARAH SHAW AND OTHERS, 12 AUG. 1771.

To The Honourable William Tyron Esqr. Governor In & Over
his Majesties Province of Newyork & and the teritories thereon de-
pending in America, Chancellor, And Vice Admiral of the same.
The Petition of Sarah Shaw Widow & Relict of John Shaw Late
of the City of Newyork Yeoman Deceased, that Neal Shaw William
Castle & Mary his Wife that These are the Children of Margerett
McDougall, Daughter of the sd. John & Sarah Shaw Most humbly
Shrweth

48

That About the year of One Thousand Seven hundred & forty your Petitioner Sarah Shaw together with her husband John Shaw, her Children Danl. Shaw & Mary Castle then Mary & One Christian Browne Since decd. Enfants Left there habitation, in the Shire of Arguile in Scotland in the Iland of Great Britain, & Embarked On board the Ship Happy Return Captn. Locklin, Campbell for this Port, being Encouraged by the Assurance Given by the Said Capt. Campbell, that Every head of a family Should on ther Arrival in America Should obtain a grant of a 1000. Acres of Land & that every Child that was a full Passenger should have 500. Acres Each, That your Petitioner Margt. Shaw was born at Sea in the Voyage so that the sd. John Shaw To, Gether With his Wife Children & the sd. Christian Browne Made Up Six Passengers.

That On their Arrival in Newyork they Underwent The Greatest hardships By the Land not being Granted According to the sd. Captn. Canpbells Assurances & their distress Was very Much. Heightened, As the sd. John Shaw, Nor Any One of his family Could Spake One Word of English. & the sd. John Shaw Was Obliged to work at hard Labour During the Rest of his Life for the Maintaince of himself and family & died Abt. Eleven years Since Without Obtaining Any Land at all & by his death yr. Pertitioner the sd. Sarah Shaw looks Upon herself in Right of her sd. husband to be Entitled to such Quantity of Land as he would have binn Entitled to had he been Living & the Said Christian Browne Being some time since deceased, the sd. Sarah Shaw as her Sister & hier at law, Looks Upon herself to be Entitled also To, the Sd. Lands of Christian Brown, which were to have bin Granted had the same bin Obtained in the Lifetime of the Said John Shaw, & Christian Brown.

That, your Pertitioners the sd. Neal Shaw Mary Castles & Margt. McDougall, Humbly Presume that they are Entitled To such Quantity of Land Each, as Were Originally Promised To be granted to Children of Passengers who, Came with the sd. Captn. Campbell Namely 500. Acres to Each Child. And more so, as the Sd. Danl. Shaw the Eldest Son of the said John & Sarah Shaw has Allready Obtained a Quantity of Land by Virtue of the Right Under Which your Pertitioners His Brothers & Sisters Claim.

That, abd. 8 years Since Aplication was made to your Pertitioner Sarah Shaw by One George Campbell Duncan Reade & Alexr. McKnight for money for her & childrens. Proportions of the Exspence of Surveying & Obtaining the Lands in the Argile, Patent, Which Severall Proportions the sd. Sarah Shaw did then Accordingly Pay. And has Since Chearfully Contributed to that End as often as she has binn Asked so to doe but Notwithstanding all the Exspence that she has pd. Neither your Pertitioner the said Sarah Shaw Nor Any One of your Pertitioners have Obtained Any Land, Tho, the Argyle Patent Out of which the said Lands Was to have, been Granted to your

49

APPENDIX

Pertitioners has Binn Some time Since divided & the Only Satisfaction your Pertitioners have Binn Offered Upon their Appling to the Trustees is to have So much money repaid to your Pertitioners the sd. Sarah Shaw, as she has Contributed On the Behalf of herself & the Rest of your Pertitioners her Children Which your Pertitioner Cannot think to be an adequate Satisfaction for their Writes in the said Lands.

Your Pertitioner therefore humbly Pray your Excellency To take their Case into Consideration, & that your Pertitioners May Obtain as much Land as they are Entitled To in Equal Proportions With the Rest of the Propriators of Land in The Arguile Patent. Or if it Should Appear that the whole Of the Sd. Patent Should be divided that your Pertitioners May Be Allotted So much Land in some Other Pleace as May be Equivalent to their Wrights in the Sd. Arguile Patent. and that in that Case the Trustees for the said Arguile Patent may Pay Back to the sd. Sarah Shaw as much Money as she has Already Paid in Respect of Obtaining The Lands in Said Arguile Patent.

And Your Pertitioners Will for Ever Pray.

Newyk. Augst. 12. 1771

(Addressed)   To His Excellency Willm. Tryon Esqr. Present.

(Endorsed)   Petition of Sarah Shaw Widow of John Shaw. Recd: 26th Augst. 1771.

1771 Augt. 28 Read in Council and referred to a Committee The lands were granted to the Petitioners Brother

Rejected. (New York Colonial Manuscripts, 97:73, in the New York State Library, Albany, N. Y.)

Document XII

DEED TO LOT NO. 32 OF THE ARGYLE PATENT GRANTED TO ALEXANDER (I) MCNAUGHTON

(NOTE. Alexander McNaughton was one of the five original trustees of the Argyle patent. The trustees of said patent were required to convey the lands granted to themselves as individuals to some other person in trust.

In order to comply with this provision of the patent, the lands of Alexander McNaughton were conveyed to his only surviving son John (2) McNaughton.)

This Indenture Made the 15th. day of January in the year of our Lord 1765 Between Duncan Read of the City of New York, Gentleman; Peter Middleton, of the same City, Physician; Archibald Campbell of the same City; Merchant; Alexander McNachten, of Orange County, Farmer; and Neal Gillaspie, Of Ulster County, Farmer, of the one part: And John McNachten of Orange County, Farmer, of the Other part.

50

THE ARGYLE DOCUMENTS

Whereas, his present majesty George the Third, by the Grace of God, of great Britian, France, and Ireland, King, Defender of the Faith, &c. by his certain Letters Patent, under the great seal of the Province of New York, reciting as is therein recited, did give and grant unto the said Duncan Read, one Neal Shaw, the said Archibald Campbell, Alexander McNachten, and Neal Gillaspie, and to their Heirs and Assigns, All that certain Tract or Parcel of Land, by the same Letters Patent created a Township, by the name of Argyle, situate, lying and being on the east side of Hudson's River, in the County of Albany.

Beginning at the East bank of the said River, at the South West corner of a tract of Land granted to James Bradshaw, and others, called Kingsbury and runs thence along the south bounds of said tract, East 492 chains to the South East corner thereof; and then along the East bounds of the said tract called Kingsbury, North four Chains: then East 236 Chains, then South 882 Chains to the middle of a stream of water called Batten Kill, then down the middle of said stream as it runs, including the half of said Creek or Kill, called the Batten Kill, to the East Bounds of a tract of land lately surveyed for Donald Campbell and others; then along the said East bounds of the said tract surveyed for Donald Campbell and others, North 367 Chains, to the North East corner thereof; and then along the North bounds of the same tract, West 317 Chains to the East bounds of a tract of Land granted to John Schuyler Junior, and others; then along the said East bounds of the last mentioned tract North 90° East 651 Chains to the North East corner of said tract: West 33 Chains, then South 60° West six Chains to a tract of land granted to Steven Bayard; then along the North bounds of the last mentioned tract, then West 205 Chains, to Hudson's River; then up the stream of the said River as it runs, to the place where this tract first began; containing 47,450 acres of land, with the usual allowance for highways, together with the hereditaments and appurtenances thereunto belonging.

To have and to hold the same tract of land and premises, with the appurtenances thereby granted and confirmed (except as therein is excepted) unto them the said Duncan Read, Neal Shaw, Archibald Campbell, Alexander McNachten and Neal Gillaspie, their heirs and assgns, for ever, to, for and upon the several and respective use and uses, itents and purposes therein expressed, limited, declared, and appointed, of and concerning the same, and every part and parcel thereoff, subject to such Quit-rents, reservations, and restriction as in and by the same Letters Patent are reserved, limited and declared, of and concerning the same tract of land and primises.

And in and by the same letters patent, the said Duncan Read, Neal Shaw, Archibald Campbell, Alexander McNachten, and Neal Gillaspie, their heirs and assigns, are particularly authorized and directed, to cause the said tract to be divided among those for whose use the

APPENDIX

same is thereby granted; and to release their respective shares thereof
to them, and in all things to execute the said trust in such manner and
form as by the same letters patent is prescribed and directed, as in
and by the same Leters Patent, recorded in the Secretary's Office of
the Province of New York, in Libro Patents, No, 14, Pages 3 to 17,
among other things therein contained, may more fully appear, relation
thereunto being had.

And whereas, the said Duncan Read, Neal Shaw, Archibald Campbell, Alexander McNachten and Neal Gillaspie, the Trustees in the
said Letters Patent named, agreeable to the directions thereof, for the
equitable locating in the said township of Argyle, the situation and
place of each of the smaller lots or tracts, for which the uses and
trusts are therein respectively limited and declared, after public notice
given to the persons for whose uses respectively the said Township
was so granted to them in trust, have caused lots to be drawn by
Ballot, for the place where the several and respective quantities of land,
so holden for them respectively in trust as afore said, should, in the
said Township, be located and fixed; and have likewise caused the
said Township to be actyally surveyed and divided, and the several
shares and allotments to be measured out, for the respective persons
for whom they hold in trust, in the several places in the said Township, whereon by the balloting aforesaid they were fixed and
ascertained.

and in order to divide the said Township to the best advantage of
the parties interested therein, the said Trustees have likewise caused
part thereof to be laid out in town lots, and the residue thereof in
convenient farms; and for the better distinguishing of said Lots and
Farms, have caused the said Town Lots to be marked and distinguished
by numbers, from the Number One, to the Number 141, both inclusive; and the said Farms, from the like Number One, to the Number
141, both also inclusive; the share appropriated by the same Letters
Patent for the use of a Minister and School-Master, being first set
apart and ascertained for those purposes; and have likewise caused a
Map or Plan of the entire subdivision of the said Township to be
made and subscribed by Archibald Campbell of Rariton in New Jersy,
and Christopher Yates of Schenectaby, the Surveyors, who run out
and surveyed the same, to be preserved as a testimonial of the execution of that part of the trust reposed in them the said Trustees, in
and by the Letters Patent (as by the same Map or Plan will fully
appear, relation being being likewise thereunto had).

And whereas, since the said several proceedings toward the execution of the said trust, he the said Peter Middleton, party to these
presents by force and virtue of certain Indentures of Lease and Release, beari..g the date the 12th & 13th of October last past, and
made between the aforenamed Neal Shaw of the one part, and him,
the said Peter Middleton, of the other part, is in due form of Law

52

become one of the Trustees of and concerning the said Township of Argyle, for the persons interested therein, in the place and stead of him, the said Neal Shaw, according to the meaning, form and effect of the same Letters Patent (as by the same Indenture of Lease and Release, may fully appear, relation being thereunto had).

And whereas, all and singular the lands and premises hereafter in these presents particularly mentioned and described part of the township of Argyle, are, by virtue of the said Letters Patent, and the several proceedings of the said Trustees, in the execution of the trust thereby reposed in them, become the distinct and separate right and property of the aforesain John McNachten party to these presents, who by virtue of misne conveyances is become legally entitled to the right and share of his father Alexander McNachten, for whose sole and seperate use 600 acres and the usual allowance for highways, part of the said township of Argyle are especially granted, limited and appointed in and by the said Letters Patent.

Now therefore this Indenture Witnesses, that they, the said Duncan Read, Peter Middleton, Archibald Campbell, Alexander McNachton and Neal Gillaspie for accomplishing the trust reposed in them by the said recited Letters Patent and also in consideration of the sum of Ten Shillings, Crrent Money of the colony of New York, to them in hand paid by the said John McNachton at or before the ensealing and delivery of these presents, receipe whereof we do hereby acknowledge, have granted, assigned, released and confirmed, and by these presents do grant, assign, release and confirm unto the said John McNachton in actual possession now being, by virtue of a bargain, sale and lease for one year, to him thereof made by the said Duncan Read, Peter Middleton, Archibald Campbell, Alexander McNachton and Neal Gillaspie by Indenture, bearing date the day before the date hereof; and also by force of the statute for transfering of uses into possession and to his heirs and assigns for ever. All certain Lot of Ground in the town Plat of the said Township of Argyle, distinguished by No. 32 of the Town Lots, bounded as follows to Wit— Begining on the South Side of the Street at a Beach Tree marked 31 & 32, then running East 13 chains and 72 links to a Walnut tree marked 32 & 33, then running South 43 chains and 73 links to a Maple Sapling marked 32 & 33, then running West 13 chains and 72 Links to a stake 15 links East from a Beach Sapling marked 32 & 31, then running North to where it began, containing 60 acres, strict measure; & Also all that Farm in the said Township distinguished by No. 32 of the farm lots Bounded as follows to Wit Beginning at a stake south east 14 links from an Elm Tree marked 31 & 32 then running South along Campbells land 49 chains 57 links to a black Oake tree marked 32 & 33, then running East 115 chains & nine links to its corner in the middle of a brook, & a Basswood tree standing on the West side of the said brook notched for the said corner, then

North 49 chains 57 links to a Beach Saplin marked 32 & 31 thence West to where it began, containing 570 acres, including the usual allowance for highways Together with all and singular the Profits, Privileges, Advantages, Emoluments, Rights, Members, Hereditaments, and appurtenances to the same hereby released premises belonging, or in any wise appertaining; and all the Estate, Right, Title, Interest, Reversion, Claim and Demand whatsoever, of them the said Duncan Read, Peter Middleton, Archibald Campbell, Alexander McNachton and Neal Gillaspie, of, in, and to the same: To have and to hold the same premises hereby granted and released, or mentioned, or intended so to be, with their, and every of their Appurtenances unto the said John McNachten, Heirs and Assigns, to the only proper use and behoof of him the said John McNachten, Heirs and Assigns forever, in as full and ample manner, to all intents and purposes, as they, they said Duncan Read, Peter Middleton, Archibald Campbell, Alexander McNachten and Neal Gillaspie, may, can, or ought to hold or grant the same, by force and virtue of the same recited Letters Patent; and subject to such Quit Rents, Reservations and Restrictions, as are in the same Letters Pattent mentioned, reserved and limited, of and concerning the same hereby granted Premises.

And They, the said Duncan Read, Peter Middleton, Archibald Campbell, Alexander McNachten, and Neal Gillaspie, for themselves, their heirs, executors and administrators, Do covenant, grant and agree to and with the said John McNachton, Heirs, and Assigns, by these presents, that for and notwithstanding any act, matter or thing, done, committed, or suffered by them, or either of them, the same hereby granted premises are free, and clear from all inumbrances whatsoever.

In Witness whereof, the parties to these presents have hereunto interchangeably set their hands and seals, the day and year first above written.

Seal'd and delivered in
the presence of us,
Angus Read.
Alexander?

The above is a printed parchment. On the back of the same parchment is as follows:

1st (In the hand writing apparently of Alexander McNachten)

"Know all men by these presents, that I, John McNachten, for and in consideration of One Hundred Pounds, York Money and for divers other good causes and considerations me thereunto moving, and acknowledging the receipt, have by these presents assigned, set over, and by these presents do assign, set over, and deliver unto Alexander McNachten, my father, all that of the within Indenture or Release for the Land Right therein mentioned, Isay assigned, set over, and delivered, from me for ever to him, the said Alexander McNachten, his heirs and assigns for ever, as may be seen by a Deed of Sale bearing even date with these presents. As witness my hand, this 22 day of April. 1765.

		his	
Signed sealed ans delivered	John	X	McNaughten.
in the presence of us.		mark	
Will: Cairns			
Duncan Gilchrist.			

2nd. In a handsome autograph, a deed of the same premises from John McNachten to Alexander McNachten, for and in consideration of the sum of £100, dated Sept. 10th. 1766. Sealed and delivered in the presence of Daniel Johnson, and John Mc Kesson

3rd. "Be it remembered that heretofore, that is to say, A. D. 1782, or thereabouts, there was a certain instrument of conveyence executed on the within written deed, declairing that Alexander McNauchten, Senior, mentioned in the deed above, for the consideration of the payment of sundry debts dues and demands justly due and owing by the said Alexander McNaughton to sundry different persons, and also in consideration of extending a filial duty and care to John Mc-Naughton, the son of the said Alexander McNaughton, for divers other good causes and considerations him thereunto moving, did remise, release and forever quit claim all his right, title, interest, possession, claim, and demand, of, in, and to the tract of land, viz. Lot No. 32, it being one of the Farm Lots of the town of Argyle in the County of Washington and the State of New York, to Alexander McNaughton junior, and Archibald McNaughton, Grandsons of the said Alexander McNaughton, senior, and sons of the said John McNaughton, to have and to hold the said lot or tract of land and premises above mentioned, unto the said Alexander McNaughton, junior and Archibald McNaughton, their heirs and assigns forever and since the execution of the said, instrument, it appears that the said instrument is partly defaced, obliterated, and worn out, and in order to substantiate, fulfil, perpetuate and perfect the agreement, intention, design and meaning of the said parties, John McKnight of the said town of Argyle and William Robertson of the same place, two of the subscribing witnesses of the original instrument of conveyance mentioned to which this alludes, personally appeared before me Ebenezer Russell, Esq. first judge of the Court of Common Pleas for the County of Washington, and being duly sworn, depose and declare, that they saw

APPENDIX

Alexander McNaughton senior, mentioned in this memorandum Execute the instrument herein alluded to as his free and voluntary act and deed for the purposes therein and herein mentioned, to the said Alexander McNaughton, junior, and Archibald McNaughton, as his free and voluntary act and deed for the uses and purposes mentioned, that they severally signed and subscribed their names as witnesses to the execution of the same in the presence of each other, and having examined the same, this 8th. day of July 1793, do allow it to be recorded. (signed) Ebenezer Russell.

4th. A quitclaim deed, from Archibald McNaughton to Alexander McNaughton, dated January 8th. 1818, recorded in Libro N. of Deeds, page 64.

These McNaughton papers were copied November 4th. 1847, from the original documents, then in the possession of Samuel Dobbin, who at that date was the owner of lot No. 32. He married Anna (4) McNaughton (Alexander 3, John 2 Alexander 1. The latter was the original grantee of lot 32 of the Algyle patent.)

THE TURNER PATENT

August 7, 1764, there was granted by the Crown to Alexander Turner and twenty-four other citizens of Pelham, Mass., 25,000 acres in what was later the town of Salem, Washington County, New York. A list of the grantees furnished by the New York State Library follows:

Alexander Turner

James Turner	Hugh Bolton	Benjamin Southwick
Thomas Johnson	James Lukes	Daniel Ballard
Matthew Bolton	George Thompson	Samuel Southwick
John McCreles	Jonathan Marsh	Daniel McCollem
John Crawford	William Crossett	Joshua Conkey
John Lucore	Alexander Turner, Jr.	William Edgar
Robert Hamilton	Joseph Rugg	William Conkey
Charles Kidd	Thomas White	Adam Clark Grey

WASHINGTON COUNTY
FAMILIES

THE MCNAUGHTON FAMILY

The McNaughton family is connected by blood and marriage with so many of the families of the Somonauk Church, and the history of the American head of this house is so woven into the history of the Argyle Patent that, although the family remained in Washington County, New York, their lineage is indispensable to this volume. The following account of the family is mainly from a series of articles written by the Honorable James Gibson and published in *The Salem Review Press* in 1887.

"The history of this family shows the great antiquity of its origin, and in many particulars is exceedingly romantic.

"It originated in Argyllshire, Scotland, and its principal seats have been located in the highlands of that section, and from thence immigrated all the early settlers of the name who came to this part of America. When we consider the wonderful tenacity with which the Highlander holds fast to the names used in the family and find there is no ancient family of McNaughtons in this section that has not among its children the names of Alexander, John, Malcolm, Donald, Daniel or Duncan, we should expect to find the same names among their ancestors in Argyllshire. Accordingly, turning to the pages of history of the Scottish clans: The NECTHAN'S, as the name was called by the Keltic race, existed and were powerful long before the introduction of surnames among them. The heads of this clan were for ages Thanes of Loch Tay, and possessed all the country between the South side of Loch Fyne and Lochawe. (*Buchanan's History of the Origin of the Clans*, p. 84.) 'Later Donald McNaughton, of Argyllshire, nearly connected with the McDougalls of Lorn joined his clan with that of the former against Robert the Bruce in the great battle of Dalre, A. D. 1306. His son and successor, Duncan, was a loyal subject of King David II, who as a reward for his fidelity conferred on his son Alexander lands in the island of Lewis which the clans long held, and the ruins of their castle on that island are still pointed out.' (*Anderson's Scottish Nation.*)

"Donald, a younger son of the family, was, in 1436, elected Bishop of Dunkeld. Alexander, of that ilk, who lived in the beginning of the 16th century was knighted by James IV, whom he accompanied to Flodden and in that disastrous battle lost his life. His son John

57

was succeeded by his second son Malcolm who died near the end of the reign of James VI, and was succeeded by his eldest son Alexander. John, the latter's grandson was with his clan under Claverhouse, at the battle of Killiecrankie, and largely contributed to its favorable result.

"It is thus seen that few families in Washington county can trace a more ancient lineage than the McNaughtons, as it can readily be followed back for more than eight hundred years. . . .

"Alexander McNaughton was the first settler of the name in this county, which he always wrote Alexander M'Nacthen—this is "Alexander, the Son of Nacthen,' that being the family name of the race, traced back, as was done in the opening section of this sketch, for more than eight hundred years. He was born in Argyllshire, in Isla, the most southern island of the Hebrides and immigrated in the first company brought over by Laughlin Campbell, landing in the city of New York, in July, 1738. He brought with him his wife [Mary McDonald] and children John, Moses, Jeannett and Eleanor. Not obtaining the promised grant of lands on which to settle in this county, he and family, with many others of his associate colonists, settled at [Tappan] in Orange county, and there remained till his removal in 1765 to the Argyle patent.

"Alexander McNaughton left a brother Duncan in Scotland, who had married Margaret, a sister of Donald Fisher, who had become the owner of some of the military patents located in Pawlet and Hebron and perhaps on his invitation she came to America, her husband having died in Scotland, bringing with her all the children she had, except Malcolm, who had come before, and was with his uncle Alexander on the Argyle patent or subsequently came there with him.

"In the grant of the Argyle patent as finally made in 1764, a trust was created for the benefit of all of the settlers who came to this country in the three companies brought over by Laughlin Campbell in 1738, 1739 and 1740, of the descendants of such of them as had died, or those of their families surviving. In this trust Alexander McNaughton was the presiding trustee and the affairs and management of the trust were largely under his direction. In order to provide for the expenses of the surveying and allottment of the lands, an assessment was made according to the number of acres allotted, and on receiving his deed the grantee would pay his share of the expenses. But as some of the parties or immigrants entitled to shares never came forward to receive their deeds and pay their portion of the expenses, all such shares were sold and conveyances made to the purchasers. In this way persons not of the original immigrants, became owners of shares in the Argyle patent. And indeed there were cases where the conveyance was made, and the expenses paid, but the grantee never claimed or occupied the lands and those who did actually occupy, had possession without any title.

WASHINGTON COUNTY FAMILIES

"In this connection an explanation may be made of how the patent received the name of Argyle. The common statement that it was originally granted to the Duke of Argyle and that he parcelled it out among his clan, is withiut the slightest foundation. The Duke of Argyle had nothing whatever to do with the grant of the patent, or with its allottment or settlement. The whole subject is matter of history and it is difficult to see how such a story could have originated. The learned and distinguished Dr. Asa Fitch, now deceased, exploded this fable more than forty years ago. (See Fitch's *History of Washington County, New York*, Section 78.)

"Laughlin Campbell, a native of Isla, which forms part of Argyllshire, in Scotland, had received encouragement from the Provincial authorities of New York, that if he would procure the immigration to the province of a number of families from Scotland, those brought over by him should receive a grant of lands free of expense sufficient to enable them to obtain a support. The object of the government of New York, in this matter, was to procure the settlement of that portion of this county lying south of what is now Whitehall, and on the borders of Wood creek, and form a barrier against French and Indian invasion from Canada by way of Lake Champlain. In pursuance of this encouragement, Campbell procured the immigration in 1738, of a colony from Argyllshire consisting of 33 families and 49 single persons, making in all 177 persons. In 1739, he in like manner, procured an immigration of 42 families and 24 single persons, making in all 193 persons. And in 1740, he obtained 15 families and 46 single persons in addition, making together 100 persons. The immigration having been obtained, all solicitude on the part of the provincial authorities to fulfill the promises made to Campbell in their behalf, wholly ceased, and no grant of lands for their settlement was made and they were left to take care of themselves as best thy could.

"The colonists thus introduced, suffered great hardships for many years, and this seems to have finally shamed those having control to make the grant of lands as originally promised. In the meantime, the lands about Whitehall and Wood Creek had been granted to others, and were included in the Skenesborough and Artillery patents, and could not, therefore, be granted to the Scotch settlers. The lands in the Argyle patent were therefore granted in their place. Thus, after the lapse of over twenty years, the settlers, or their descendants, who came over under the offers made to Laughlin Campbell, received a grant of those lands in part fulfillment of the original promises made to him. This grant was made by the Governor and Council of the Province of New York, by patent to Alexander McNaughton and others, in trust, to be allotted among those settlers and their descendants. The patent was issued in the usual form of such grants and in the same form and manner as Skenesborough and other patents located in this section were issued.

59

APPENDIX

"This much for the story of the Duke of Argyle granting or receiving a grant of the patent. The name of Argyle was given because the settlers were all from the Shire of Argyle in Scotland. "Alexander McNaughton settled on that portion of the patent which now lies in the town of Greenwich, and on the farm which was long after occupied by Deacon Samuel Dobbin as a homestead. Here he built a common log house in 1764, and a few years after another of squared logs. He was appointed a justice of the peace, and was the first one appointed on the Argyle patent to that office.

"It was while acting as such justice that he was summoned to New Perth, as Salem was then called, to enforce the law against Ethan Allen and his ruffianly associates, who had by force of arms raided the lands granted to Charles Hutchan, Donald Campbell and others in the northeast corner of the present town of Salem and had torn off the roofs from their log houses, and by threats compelled the occupants to leave the premises."

The fore going petitions and memorials that resulted in the granting of the Argyle Patent have made plain the large credit that is due Alexander McNaughton for his sagacity in acting for the colonists and for his subsequent administration of the trust created for the benefit of the settlers brought by Laughlin Campbell, in which trust he was the presiding trustee and principal administrator.

ALEXANDER (1) McNAUGHTON, one of the five original trustees of the Argyle Patent, was born in the Isle of Islay, Scotland, about 1693; died in the home of his son-in-law, Hon. Edward Savage, in Salem, New York, in 1784; married in Argyleshire, Scotland, about 1725, Mary McDonald, born there in 1690; died in the home of Duncan Taylor, a relative, where the family had tarried on their way home from Burgoyne's camp.

Alexander McNaughton, with his wife and their four elder children, came with the first of Captain Lauchlan Campbell's Highland Scotch colonists. They left Scotland in July 1738 and landed in New York September 22. Settling first on the Kakiate Patent—the name of a patent, not the name of a town or township—they later removed to Tappan in Orange (now Rockland) County, New York, where they resided when the Argyle Patent was granted in 1764. The next year this family and the Livingston family settled on the Argyle Patent.

Children:
    i. John (2) born on the island of Islay about 1726; died before 1800, in the McNaughton homestead in Greenwich, Washington County, N. Y.; married about 1752, Margaret, daughter of Duncan and Mary (Gillis) Taylor of Argyle.
    Children:
    Alexander (3).
    Archibald.

60

Mary, married James Mains.
Robert.
Daniel.
Eleanor, married Col. John McCrea, a brother of Jane McCrea.
Margaret, married David McKnight.
ii. Moses, died aged about twenty-one years in Orange County, N. Y. He was the schoolmaster in the family and taught the other children, under the supervision of his mother.
iii. Janet, married Archibald Brown and died in Argyle June 22, 1770. Her remains were the first interred in the old Argyle cemetery, the land occupied by the cemetery being a part of her husband's farm. They had no children but had taken her niece Janet (Jane) (3) Livingston to live with them, who was three years of age at her aunt's death.
iv. Eleanor, born May 5, 1735, in the Island of Islay; died in the home of her daughter, Mrs. James Shaw, in East Greenwich, N. Y., Mar. 7, 1817; married in Tappan, N. Y., Nov. 23, 1756, Archibald (1) Livingston, later owner of Lot No. 66 in the Argyle Patent, N. Y. See page 352.)
v. Mary, born in Orange, N. Y., Apr. 24, 1742; died in Salem, N. Y., Feb. 23, 1834; married in Salem, Dec. 31, 1767, Hon. Edward Savage. (See page 352.)
A half mile south of the village of Argyle, New York, is situated the old cemetery where lie the unmarked graves of some of the earliest of the colonists. In memory of these pioneers a bronze tablet was erected here by James A. and Henry J. Patten, which was dedicated in June, 1923. The inscription reads:

In this cemetery are interred the mortal remains of

MARY McDONALD, WIFE OF ALEXANDER McNAUGHTON. Born in Argyleshire, Scotland, in 1690. Died in Argyle, N. Y., in 1777.

JANET McNAUGHTON, her daughter. Wife of Archibald Brown. Died in Argyle, N. Y., June 22, 1770..

MARY LIVINGSTON ROBERTSON, her granddaughter.

WILLIAM PATTEN. Born near Stonebridge, Ireland, November 5, 1752. Died in Argyle, N. Y., December 12, 1841.

MARTHA NESBITT, his wife. Born in Kilmore, Ireland, 1752. Died in Argyle, N. Y., March 2, 1817.

The land occupied by this cemetery was formerly the homestead of Archibald Brown and the remains of his wife were the first interred in the cemetery.

Alexander McNaughton was born in Argyleshire, Scotland, about 1692. Died in Salem, N. Y., in 1784. His remains were interred in

61

the McNaughton burial ground on his own land, lot 32 of the Argyle patent.

Eleanor McNaughton, his daughter, wife of Archibald Livingston, was born in the Isle of Islay, Scotland, May 5, 1735. Died in East Greenwich, N. Y., March 7, 1817.

Archibald Livingston, her husband, was born in Argyleshire, Scotland, in 1730. Died near East Greenwich, N. Y., September 2, 1792.

Mary Livingston, their daughter, was born in Tappan, New York, September 26, 1757. Died in Argyle, N. Y., August 7, 1793.

William Robertson, her husband, was born in Peterhead, Scotland, January 19, 1752. Died in Argyle, N. Y., February 15, 1825.

Mary Robertson, their daughter, was born in Argyle, N. Y., August 7, 1793. Died near Sandwich, Ill., April 6, 1890. James Patten, her husband, was born near Stonebridge, Ireland, July 4, 1793. Died in Salem, N. Y., December 21, 1827. Son of William Patten and Martha Nesbitt.

THE LIVINGSTON FAMILY

ARCHIBALD (1) LIVINGSTON born in Argyleshire, Scotland, in 1730, died in his home in Argyle, New York, Sept. 2, 1792. With his parents he went to the north of Ireland in 1744, came to America in 1751, settled among the Highland Scotch in Orange County, N. Y., removing to Washington County in 1765. He married at Tappan, N. Y., Nov. 23, 1756, Eleanor (2) a daughter of Alexander McNaughton, born May 5, 1735, in the Island of Islay. Archibald Livingston became the owner of Lot No. 66 in the Argyle Patent.

Children:

    i. Mary (2), born Sept. 26, 1757; died in Argyle, N. Y., Aug. 7, 1793; married Sept. 24, 1775, William (3) Robertson. (See page 227.)

    ii. Margaret, born May 30, 1759; died in Argyle Dec. 7, 1839; married about 1783, John Taylor born 1748; died Apr. 16, 1813; son of Duncan and Mary (Gillis) Taylor, of Argyle.

    iii. Janet (Jane), born Feb. 2, 1767; died in Cambridge, N. Y., Feb. 20, 1853; married first, in Argyle, Aug. 21, 1800, James Shaw; born in the parish of Kilmadock, Perthshire, Scotland, in 1768; died near East Greenwich, N. Y., Nov. 24, 1822. Son of John and Margaret (Thompson) Shaw.

       Mrs. Shaw married second, in East Greenwich, N. Y., May 16, 1826, William Stevenson, born in the parish of Steinkirk, Galloway, Scotland, Feb. 15, 1772; died in Coila, Washington County, N. Y., July 8, 1844.

WASHINGTON COUNTY FAMILIES

iv. Hon. Alexander, born June 8, 1769; died Oct. 23, 1863; married, 1806, Elizabeth (2) McDougall; born in Argyle (now Greenwich) in 1787; died on the Livingston homestead, Lot. No. 66 of the Argyle Patent, Feb. 28, 1853. She was a daughter of William and Sarah (Gilleland) McDougall. William was a soldier of the Revolution.

v. Moses, born Mar. 2, 1772; died Aug. 24, 1793.

vi. Marianne, born June 29, 1774; died near East Greenwich, N. Y., Feb. 12, 1842; married in the home of her father, Apr. 7, 1801, Alexander Shaw, born in 1764, son of John and Margaret (Thompson) Shaw.

vii. Eleanor, born in Argyle, N. Y., Aug. 10, 1777; died near East Greenwich, N. Y., Apr. 24, 1855; married in the home of her father, Aug. 6, 1798, William (2) McDougall, Jr., born in New York City Sept. 23, 1770; died near East Greenwich, N. Y., June 17, 1819, son of William (1) and Sarah (Gilleland) McDougall.

Mrs. Archibald Livingston (Eleanor (2) McNaughton) told her granddaughter, Mrs. James (4) Patten, who spent the first twenty-four years of her life with her grandmother, that nearly all of the first Highland Scotch settlers on the Argyle Patent were related either by blood or by marriage. Mrs. Livingston also said that there were Campbells among the Argyle colonists who were related to the Duke of Argyle. Being ten years of age at the time of the Rebellion of 1745 she remembered it distinctly and narrated to her descendants many tales of Bonny Prince Charley.

The torch of tradition lighted by Eleanor Livingston and handed on by Mrs. Patten kindled in her granddaughter, Jennie M. Patten, the interest that culminated in the church history.

THE SAVAGE FAMILY

The Savage family is of French origin. They were driven from France by the revocation of the Edict of Nantes in 1685. They settled for a time in the north of Ireland, members of the family intermarrying with persons of Scottish descent. A portion of the family came to America in 1716 and settled in Massachusetts.

JOHN (1) SAVAGE, born in 1706, was appointed captain of a company of volunteers in 1758 and served under General Bradstreet in his expedition against Fort Frontenac and under General Abercrombie in the assault of Fort Ticonderoga. He moved to Salem in 1767; married Eleanor Hamilton and died there in 1792, aged 85.

EDWARD (2) SAVAGE, a son of John, born in Rutland, Mass., January 9, 1745, came to Salem with the family in 1767. He was

63

the first sheriff of the county after the Revolutionary War; also surrogate. A member of the state legislature for twenty-one years, he was three times elected a member of the council of appointment. He took part in the battle of Plattsburg in 1814, and died October 13, 1833, aged 87. Married December 31, 1767, Mary (2) a daughter of Alexander McNaughton, born in Orange County, New York, April 24, 1742; died in Salem, New York, February 23, 1834.

Children:
- i. Alexander (3), died in infancy.
- ii. Jane, born July 6, 1777; died Jan. 27, 1802; married, 1800, Rev. Joseph Sweetman. Child: Jane Sweetman, married Rev. Nathaniel Bacon, of Niles, Mich.
- iii. John, born Feb. 22, 1779.
- iv. Mary, born Nov. 22, 1782; died Apr. 29, 1784.

JOHN (3) SAVAGE, LL.D., born February 22, 1779, in Salem, New York; died in Utica, Oct. 19, 1863; married, Feb. 27, 1810, Esther, daughter of Gen. Timothy Newell, who died Mar. 14, 1811. Was graduated from Union College in 1799, receiving first honors, and was admitted to the bar. Appointed Chief Justice of the Supreme Court in 1822, he held that office until 1836, when he resigned. He married second, Ruth Wheeler, of Lanesboro, Mass., Nov. 8, 1816.

Children:
- i. Mary Ann (4), born Apr. 1, 1819; died May 18, 1846; married Nov. 8, 1837, Hon. Ward Hunt.
- ii. Laura Wheeler, born Oct. 28, 1822; died March 2, 1905.

THE GILLASPIE FAMILY

NEAL (1) GILLASPIE, married Mary McIlpheder and with two oldest sons came with Captain Lachlin Campbell in 1739. Neal Gillaspie was one of the original five trustees of the Argyle patent and a relative of the McNaughtons.

Children:
- i. Gilbert (2).
- ii. Angus.
- iii. Daniel, married and had a son Gilbert.
- iv. Catharine, married William Goodson.
- v. Neil, married Mary Van Winkle.
  Children:
  - i. Catharine (3), married John Winne.
  - ii. Mary, married Casparus Bain.
  - iii. Eleanor, married George Ferguson.
  - iv. Janet, died unmarried.
  - v. Nancy, died unmarried.

vi. Margaret, died unmarried.
vii. Neil.
viii. John.
ix. Jacob, married Miss Raney.
x. Daniel, married at Massena Springs, N. Y.

The will of Neal Gillaspie of the Precinct of Wallkill, Ulster County, New York, yeoman, date March 4, 1769, probated March 31, 1769.

I Neal Gillaspie of the Precinct of Wall-kill Ulster County, yeoman, being sick . . .

Whereas I was proprietor in a patent of land called the Scotch patent or Argyle patent and one of the trustees of the same whereof on the said patent, I was obliged with the other trustees to convey my part of said patent to some one it Trust. I confided to my son Neal Gillaspie and he is now vested with the deed, dated Jan. 15, 1765, and executed by myself, Duncan Reed, Peter Middleton Arch. Campbell and Alexander McNachten, trustees in said patent, of 453 acres.

I will that my son Neal shall make over by deed to my son Daniel 100 acres. To my Wife 100 acres. To my Daughter Cachy or (Cattie) wife of William Goodson (or Goodjen) 100 acres.

And the lot of 45 acres laid out for a town lot in said patent, my son Neal shall by deed of trust make over to someone for my Grandson Gilbert son of Daniel. If my son Neal does this then I leave him his share of my property as hereafter mentioned (1) My farm where I now Dwell with all the utensils to be sold by my executors and all debts to be paid and also the debts of my son Daniel. From the remainder, one third to be paid to my wife (and she is to pay 20 to my daughter Cashy wife of William Goodson,) one third to my son Daniel and one third to my son Neal.

I leave to my sons Daniel and Neal and my daughter Cashy certain cattle. I leave to my wife and children each their wearing apparel.

Whereas I perchased lot 62 in said Scotch patent of Mary Beatoy for 60 and whereas my kinsman, Alexander Campbell came to this country upon encouragement given him by me, I leave him all of said lot of 300 acres and he is to pay the 60 with interest in Seven years.

I make my wife and my good friend Alexander Kidd and David Jager Executors and my trusted friend Cadwallader Colden Jr. overseer.

Witnesses { Thomas Beatty.
Samuel Haines.
Archibald McNeal.

THE CLARK FAMILY

In the picturesque cemetery at Cedar Springs, Abbeville, South Carolina, is a tomb bearing an inscription which, though unimposing and even inaccurate in some particulars, serves to call the attention of the passerby

65

APPENDIX

to one of the most unusual and interesting characters on the pages of church history in America—the clergyman, physician, financier Thomas Clark. The inscription reads as follows:

TO THE MEMORY
OF
REV. THOMAS CLARK, D.D.
WHO WAS BORN IN IRELAND*
LICENSED TO PREACH APRIL, 1748
LABORED IN BALLIBAY 16 YEARS
EMIGRATED TO NEW YORK
28TH JULY 1764
AFTER LABORING THERE MANY YEARS
CAME TO ABBEVILLE, S. C. 1786
WHERE HE LABORED AS THE FOUNDER
AND FIRST PASTOR AT CEDAR SPRINGS
AND LONG CANE UNTIL HIS DEATH
DEC. 26TH. 1792

A member of the Clark family contributes the inscription on the graves of Mrs. Clark and their infant son, who were buried beside the church in Cahans, Ireland:

HERE LIES THE CORPSE OF
ELIZABETH CLARK, ALIAS NESBITT
SPOUSE OF THE REV. THOMAS CLARK
WHO DIED DECEMBER 18, 1762
AGED 32 YEARS. A TRUE CHRISTIAN
ROBERT CLARK WHO DIED JULY 18, 1862, AGED 6 YEARS

REV. DR. THOMAS (1) CLARK, M.D., born in Galloway, Scotland, about 1722; died in Cedar Springs, South Carolina, December 26, 1792; married in Ireland about 1752 Elizabeth Nesbitt, who was born in 1730; died December 18, 1762; probably a daughter of Thomas Nesbitt,** of Drum-a-connor, who was one of the elders who signed Dr. Clark's call to become pastor of the Presbyterian Church in Cahans, four Irish miles from Ballibay, in County Monaghan, Ireland. Robert, her brother, whose wife was Nicolina Montgomery, went security in the sum of £4000 when Dr. Clark was liberated from Monaghan jail. It has been stated that Dr. Clark was a graduate of Glasgow University, but the Registrar reports that no record can be found of a degree having been conferred upon one of that name at or near 1748.

Mrs. Clark died two years before the Exodus, but her sister and brother John Nesbitt, with the latter's wife Elizabeth, emigrated with Dr. Clark's colony in 1764. Miss Nesbitt died in Washington County, New York, at the home of her nephew, the Honorable Ebenezer Clark.

*Error. **Dr. R. Nesbitt, M.D., of Sutton in Ashfield, Notts, England, is descended in this line and has the records for 500 years.

Children born in Cahans, Ireland:
   i. Ebenezer (2), born July 4, 1753; died in Argyle, New York,
      Feb. 10, 1826; married first his cousin Elizabeth, daughter
      of John and Elizabeth Nesbitt above mentioned; married
      second Mrs. Margaret McClaughry Savage, widow of James
      Savage, of Salem.
  ii. Robert, born Dec. 22, 1755; died July 18, 1762, in Cahans.
 iii. Elizabeth, born Oct. 10, 1758; married Maj. James Campbell,
      son of Duncan, the first supervisor of Argyle.
  iv. Benjamin, whose birth occurred between 1759 and 1762, for
      which period the church records have been destroyed, was a
      physician. His will, probated in Abbeville, South Carolina,
      mentions, besides his children, his brother Ebenezer, his moth-
      er-in-law, Mrs. Mary Cochran, and James Cochran, the two
      last being executors.
      Children:
        i. Elizabeth (3).   ii. Jean N.

The reader will be glad to have these genealogical gleanings that go
so far afield and are now printed for the first time, if he should ever
have the good fortune to peruse "The Salem Book" account of this
extraordinary clergyman, physician, financier, Dr. Thomas Clark, the
Leader of the Exodus of the Cahans, beside whose prison experiences
the vicissitudes of the Vicar of Wakefield seem commonplace indeed.

No less picturesque, though less harrowing than his experiences in
Ireland, are his adventures as the first pioneer preacher on the frontier
between Albany and the Canadian border, for a large volume could be
compiled dealing with his explorations in eastern New York in search
of suitable farmlands for his "imported" congregation numbering three
hundred souls, of his determination and finesse in obtaining favorable
terms of tenure, of the vigor with which he collected the rents for the
owners and himself carried the $1,500 on horseback to their agents in
New York City, of his yeoman service in the erection of the log parson-
age and first meeting house in Salem. It is pleasant to picture his congregation coming many miles on foot and on horseback through the
aisles of the forest, in such numbers that in summer services had to be
held in the open air in the wooded glen of the nearby spring, the pastor
standing under a small open tent, his Bible and Psalm book on a little
table before him, while his flock were seated on the shelving ledges of
rock that encircled the spring like an amphitheatre.

Dr. Clark married many of the ancestors of Somonauk people, among
them William Robertson and Mary Livingston. He and his son, it is
said, boarded at the Livingston home at one time.

While the first search for lands was in progress land speculators from
the South drew away a few of Dr. Clark's emigrants to the Abbeville
District of South Carolina, and the pioneer clergyman literally rode the

circuit between these widely separated branches of his church in America. After eighteen years of service in Salem, it was to this place that he was called, and here he died on December 26, 1792, a true Link With The Past, his last act being the inditing of an affectionate letter to his former congregation in Ballibay with which he had never ceased to communicate.

A leader of thought in the old world, Thomas Clark was a pioneer of civilization in the new, and by his advanced spiritual ideals, indomitable courage, enterprise and strong personality hastened the settlement of a large area in Washington County and through the descendants of his church members contributed to the spiritual heritage of the Church at Somonauk.

The following is from Johnson's History of Washington County:

"During the time Dr. Clark remained in Salem, the amount of labor he performed is simply marvelous.

"No other than an iron constitution could have borne it. Until the arrival of Dr. John William he was the only physician in the place. In addition to his care of the church he was called to attend the sick; in addition to this he regularly visited Hebron, Argyle and Cambridge, preaching and preparing the way for the organization of flourishing churches. Like Paul, he was abundant in labors, and like his, his labors were crowned with success."

TO THE PIONEERS OF THE WEST

Would God that we, their children, were as they!
  Great-souled, brave-hearted and of dauntless will:
  Ready to dare, responsive to the still,
Compelling voice that called them night and day
To the far West where sleeping greatness lay
  Biding her time. Would God we knew the thrill
  That exquisitely tormented them, until
They stood up strong and resolute to obey.

God, make us like them, worthy of them; shake
  Our souls with great desires; our dull eyes set
On some high star whose splendid light shall wake
  Us from our dreams, and guide us from this fen
Of selfish ease won by our fathers' sweat.
  Oh, lift us up—the West has need of Men!

Mrs. Ella Higginson.